D0906911

PERSPECTIVES IN
ECOLOGICAL THEORY

RAMÓN MARGALEF

Perspectives
in
Ecological Theory

THE UNIVERSITY OF CHICAGO PRESS

CHICAGO · LONDON

Library of Congress Catalog Card Number: 68–27291

THE UNIVERSITY OF CHICAGO PRESS, CHICAGO 60637

The University of Chicago Press, Ltd., London, W.C.1

Foreword

The following pages contain the substance of four lectures delivered in the Department of Zoology of the University of Chicago, under the auspices of the Louis Block Fund, in May of 1966.

The author felt deeply honored by the invitation to come to a place so conspicuous in the history of ecology as Chicago and considered it a privilege to meet so receptive and stimulating an audience. In scientific exchange, written information is apt to be rigorous. Talks and meetings can be used for more speculative purposes, with all the advantages of instant feedback and a certain loose indulgence in anthropomorphism and analogy. Such is my treatment here as I deal with very general problems of present-day ecology. Any reference to 'new ecology' would be as uselessly pompous as the name 'new systematics' has been. My purpose is only to express the conviction that some aspects of the solid ecology of yesterday and of today allow us to build a theoretical superstructure that, perhaps, is not irrelevant after all. A certain familiarity with ecological questions is assumed from the reader, many references to my own works published elsewhere are included, and, in general, the view is held that relevant evidence does not consist of a massive accumulation of trivia.

I must acknowledge, even in Chicago, that my thinking has been strongly influenced by G. Evelyn Hutchinson and several of his former students, especially the brothers Odum.

It is an advantage for a lecturer that the printed version of a series of talks allows some flexibility and the incorporation of the results of feedback. Giving specific credit would be a difficult enterprise, since good intellectual centers are as complex as mature ecosystems. As partners in the intellectual feedback loops during my stay in Chicago, I will remember especially Robert Inger, Richard Lewontin, Monte Lloyd, and Thomas Park.

To Dr. Thomas Park, an inspiring figure for any ecologist, goes my special gratitude as the promoter of my trip; and my sense of indebtedness is extended to Dr. Burr Steinbach, Chairman of the Department of Zoology, and to the trustees of the Louis Block Fund.

The reader should not forget that this book was prepared in very bad English and has needed a lot of work from editors and colleagues to suit it for publication. The author feels deeply indebted to all the persons who have contributed to the realization of the book.

Contents

viii   *Contents*

I

The Ecosystem as a Cybernetic System

CYBERNETICS

Cybernetics concerns itself with control and communication in systems formed by living organisms and their artifacts. The word was introduced by Norbert Wiener at the time of the Second World War, resurrecting and giving new meaning to the Greek word for 'helmsman' (Wiener 1948). So much popular prestige has been attached to cybernetics that when the title of a talk is "Ecology and Cybernetics," the catchword, at least in some countries, is 'cybernetics', just as when the title of a talk is "Biology and Exobiology," exobiology is likely to draw more attention (and surely more money!) than biology. This analogy is not put forward as a malicious intimation that cybernetics is something as nearly nonexistent as exobiology, but only as a reminder that the emphasis has to be placed on the solid foundation, that is, on ecology.

Cybernetics must be considered as simply another possible way to look at things. Whether interest in cybernetics is passing or lasting depends upon whether it adds something new and valuable to the framework of general ecology. Every fashion that goes through the scientific world should leave something of value. The points of view emphasized by cybernetics are, indeed, useful in different aspects of ecology. The books of Ashby (1954, 1956), for instance, contain many stimulating suggestions of interest to the ecologist, and not only in the areas specifically discussed here.

Cybernetics refers to systems. Every system is a set of different elements or compartments or units, any one of which can exist in many different states, such that the selection of a state is influenced by the states of the other components of the system. Elements linked by reciprocal influences constitute a *feedback loop*. The loop may be *negative*, or *stabilizing*, like the one formed by a heating unit and a thermostat or the mechanisms regulating sugar level in the blood. Or the loop may be *positive*, or *disruptive*, like the spread of an annihilating epidemic.

A characteristic of negative feedback is that not only the entire system but also some selected states of the system show a considerable persistence through time. A cybernetic system influences the future, or bridges time, in the sense that the present state sets limits or patterns for future states. Thus, the present state is a bearer of information. Here we meet the word 'information' used in the context of cybernetics. Information has to do with any a posteriori restrictions of a priori probabilities. Any cybernetic system, through the interactions of its parts, restricts the immensely large numbers of a priori possible states and, in consequence, carries information. Communication or information theory is closely related to cybernetics, and may even be the same—but semantic questions do not interest us at present.

Information contained in nature—why nature is as it is and not otherwise—allows us a partial reconstruction of the past. Only a hypothetical universe composed of pure energy would be without a past. In any material system interactions and cybernetic mechanisms appear, and with them stores of information. Organisms constitute a wonderful example, but this process of history-making and history-telling is by no means restricted to the organic world. The development of the meanders in a river, the increasing complexity of the earth's crust through successive epochs of orogenesis, are information-storing devices in the same manner that genetic systems are. Moreover, all such cybernetic systems are naturally self-organizing systems. Information is expressed by a mechanism, and storing information means increasing the complexity of the mechanism. The success of life springs from miniaturization. It depends on the packing, in a small space, of a prodigious number of overlapping mechanisms, wonderfully persistent by virtue of built-in regulatory circuits and sufficiently open to carry into the future a promise of new developments.

Machines are ordinarily assembled with some finality in the mind of the constructor or designer. A machine without finality is merely an objet d'art. Organisms are objets d'art only outside their environments. In cybernetic systems that occur naturally, whether the meanders of a river or organisms, the only test to pass is the ability to remain. Finality is automatically equated with persistence, and persistence is more apparent if the external form is preserved through time. Thus, conservatism seems to be a law of nature, and systems endowed with the highest stability of form—no new properties being added that are not accounted for—may be rightly considered the best channels for information. Often, of course, the preservation of a form or a state reflects simply the unavailability of the energy necessary for a change. In any case, however, it can be deduced from general cybernetic theory that any system that can adopt different states automatically remains in, or after a time adopts, the most stable of them. It is probably unnecessary to add that we have here the basis for a welcome enlargement of the theory of natural selection.

ECOLOGY

Cybernetic systems can be recognized at many different levels in the world of life. They are found at the cellular level,

at the organismic level, and also at the level where the interacting elements are individuals. Science tends to deal only
with relations between units or elements; generalizations are
possible where elements of a class are equivalent, but not
where class boundaries are imprecise because of the heterogeneity of the members of the classes. A first step is to define
the classes. In the second step, the different behaviors of the
different classes are explained on the assumption that the
systems are differently organized but are all composed of the
same sort of more elementary entities. Such is the approach
that has been followed in the study of molecules, atoms, and
subatomic particles. A formal recognition of different levels
of organization can be useful, too, as a basis for a definition
of ecology.

Ecology, I claim, is the study of systems at a level in which
individuals or whole organisms may be considered elements
of interaction, either among themselves, or with a loosely
organized environmental matrix. Systems at this level are
named ecosystems, and ecology, of course, is the biology of
ecosystems. To explain why individuals of different species
(different classes) behave differently falls outside the scope of
ecology. This approach leaves aside much of the stuff tradi-
tionally included in books of ecology—much physiology, behavior, and physical geography. The approach, moreover,
makes unnecessary any concept of superorganism or of closed
biocenosis, from which ecology has suffered so much. In no
case do I pretend to revolutionize the table of contents of any
future textbook of ecology. My aim is only to state the context
in which I believe it is possible to speak of a theory of ecology.

FEEDBACK IN ECOSYSTEMS

A simple example of an elementary cybernetic mechanism,
in the form of a negative feedback loop, is the classical one of
a predator and its prey (Fig. 1). Organisms are the bearers of

huge amounts of information. Since they can be destroyed but cannot be produced from nothing, any regulatory mechanism implies an initial overshoot. An excessive number of offspring is produced by the prey. This number is reduced to a lower level through destruction by the predator. Such destruction is density-dependent, because the numbers of the predators themselves are dependent on the numbers of the prey at a previous time. The interactions between species can be considered cybernetic mechanisms. Their goal can be

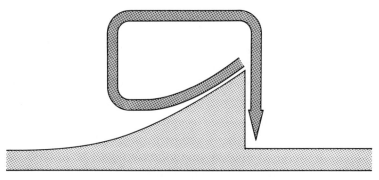

Fig. 1. Feedback loop in the regulation of numbers of a species. The excess off-spring produced is cut down to appropriate size by an agent (another species) of intensity proportional to the excess of descendants.

stated, for our own special and descriptive purposes, as the maintenance of a reasonable constancy in numbers. Such constancy in numbers means the preservation of information, inasmuch as future states of the system become more predictable from the present situation.

Although such points of view are not common to all ecologists, they may be considered implicit in ecologists' usual way of dealing with these questions, following the lead of Volterra, Lotka, and possible, if less well-known, forerunners. The way in which these authors present their theories can be summarized by writing the rate of change, dN_i/dt, of the number of individuals, N_i, of any species, and the rate of change, dE/dt, of the intensity of selected environmental factors, E, as proportional to a sum of products representing various interactions. An example might be the set of differential equations:

$$
\begin{array}{lllll}
 & E & N_1 & N_2 & N_3 \\
\hline
dE/dt = & & -aEN_1 & -bEN_2 & \\
dN_1/dt = & +eEN_1 & -fN_1{}^2 & -qN_1N_2 & -hN_1N_3 \\
dN_2/dt = & & +iN_1N_2 & -jN_2{}^2 & -kN_2N_3 \\
dN_3/dt = & & +lN_1N_3 & -mN_2N_3 & -nN_3{}^2
\end{array}
\tag{1}
$$

Although some attention has been given to the implications of similar groups of equations, ecologists have not gone very far in discussing such a model. It is true that nobody has been able to write down an expression in reasonably accurate terms, but this is an argument for passing from a 'microscopic' to a 'macroscopic' approach, as will be discussed later, and some speculative considerations on the perspectives of such a formulation are not out of place here.

Certainly, this representation needs some mending: in particular, it requires the introduction of time lags, nonlinear effects, and higher order interactions, which probably would make the whole thing forbidding for the average ecologist born in the first quarter of the twentieth century. As it stands, it is interesting to consider the matrix of the coefficients of the products of the numbers, or intensities, of interacting elements. Thus, associated with the matrix of possible cross products given in the set of equations (1),

$$
\begin{array}{llll}
E^2 & EN_1 & EN_2 & EN_3 \\
EN_1 & N_1{}^2 & N_1N_2 & N_1N_3 \\
EN_2 & N_1N_2 & N_2{}^2 & N_2N_3 \\
EN_3 & N_1N_3 & N_2N_3 & N_3{}^2
\end{array}
\tag{2}
$$

we have the matrix of coefficients of the quadratic form:

$$
\begin{array}{cccc}
0 & -a & -b & 0 \\
e & -f & -q & -h \\
0 & i & -j & -k \\
0 & l & -m & -n
\end{array}
\tag{3}
$$

According to the values of the coefficients, interactions may be more or less strong. From empirical evidence it seems that species that interact feebly with others do so with a great number of other species. Conversely, species with strong interactions are often part of a system with a small number of species having strong fluctuations.

As the rates of change represented by expressions (1) are dependent on the current values of the variables and determine the values at the next instant of time, the operations represented in expressions (1) can be visualized as an iterative process, running continuously and leading to a steady state. The steady state of the system will depend only on the values of the coefficients of interaction. More transient states will depend on changes in E and N_i not implicit in the expressions, such as changes in the physical environment or removal or inmigration of individuals of assorted species. Such changes, not given in the original formulation of the ecosystem, may induce alterations even in the values of the coefficients of interaction.

The concept of ecological niche will probably turn out to be unnecessary, allowing an always welcome simplification of ecological jargon. As the term is used commonly, a niche lumps together several species that behave in a similar way in the system. Such similarity should be deduced from the values of the coefficients of interaction of the species in question with other elements of the system. If values are identical, no ecological distinction can be made between groups, all belonging to the same species. A niche can be constructed by grouping the species for which the signs of the coefficients are the same—for example, species 2 and 3 in the expressions (1) —and for which the numerical values are not very different. Competing species have the same signs, and competition is stronger the closer the numerical values of the coefficients.

Competition thus is the result of the combination of two parallel negative feedback circuits, from which a positive feedback results (Fig. 2). Following this route around the corner, so to speak, we achieve another definition of niche: species belong to the same niche if there is some sort of a positive feedback loop between them, or, at least, one that is not negative. Except in cases of strong disoperation, such

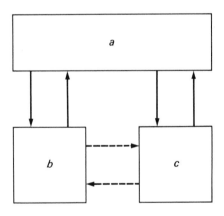

Fig. 2. Use of the same resource, *a*, by two species *b* and *c*. There exist negative or stabilizing feedback circuits between *a* and *b*, and between *a* and *c*. The result is a possible positive feedback loop between *b* and *c*.

positive feedback is the result of two parallel negative feedbacks. Both competing species use the same resource, or are used by the same piece of the ecosystem. If, after all, it is desirable to have a definition of niche, this one is perhaps the only operational one.

ENERGY AND ENERGY GATES

Perhaps the adoption of a cybernetic point of view can help to emphasize a very important and often neglected point. The two elements which are linked in a feedback loop are not equivalent; there is a basic asymmetry between them. To put it in a nutshell, when a predator kills and eats prey, the prey retains no memory of the event, but the predator receives the energy and is satisfied. One of the elements gives a surplus of energy, and the other uses a small part of such energy to maintain a more stable internal state. In other words, each interaction represented by an intersection in expressions (1) is a node of energy transfer, or, more precisely, a *gate* for energy. And as energy flows, some information—redundant information, it is to be hoped—is destroyed.

Returning to expressions (1), we see that in a steady state where $dN/dt = 0$, the total of the positive terms in every equation equals the total of the negative terms. But the absolute values of the rates of change on either side of the equilibrium may be different for different species. In species that have strong interactions (such as dominant species, weeds, pests, and so on) the sum of the components, either positive or negative, is very large. For species with low or feeble interactions, although there may be interactions with many species, the total sum is small. In ecology it is usual to include all factors that increase dN/dt in a rate b and all the factors that decrease it in a rate m. Then the differential equations in (1) typically have a solution of the form

$$N_t = N_0 e^{(b-m)t}.$$

If $b = m$, the population size N is in a steady state. Independently of such a condition, the absolute values of b and m matter very much. They represent the flux of energy, or the cost of operating the cybernetic mechanism. An organism with a high fecundity *b* and a high mortality *m* may be compared to a thermostat that uses a good deal of thermal energy to work as a regulator. Of course, it is more expensive to operate and it is a poorer regulator because it tends to produce strong fluctuations. The only positive quality of such a piece of equipment would be its ability to work under a wider range of conditions: it would be expected to be rather rugged and resistant to external fluctuations.

Ecologists using the equations of Lotka (1956) and Volterra (1926) have not emphasized their asymmetry, probably because they accepted that assymmetry as trivial. The energetic asymmetry is worth further consideration, however, because it is possible to associate some notions related to entropy with the action of such energy gates. Entropy production should be greater if the transfer of energy is higher. Moreover, some information is destroyed in the functioning of the energy gates, at least in the form of the organization and memory of the prey that are killed.

The intersections in any set of expressions like (1) are also places where entropy is being produced. Notice that every pair of species appears to be associated at two points symmetrical in position but different in relation to entropy. That is, matrix (2) is symmetric, but matrix (3) is not. Species strongly interacting (eating too much, using food inefficiently, producing too many offspring; e.g., species in sand dunes, in plankton, and in other similar ecosystems) represent sites of a rather high production of entropy.

If one were able to draw up an adequate chart for every ecosystem on the model of expressions (1), one could probably deduce the relative amount of the total entropy produced in the maintenance of the system from the matrix of coefficients. Let us suppose that the total biomass is similar in two cases. If one of the two systems under comparison contains more species and the interactions between them are, on the average, associated with weaker exchanges of energy and a smaller destruction of information, then it could be said that in this system the same increase in entropy drives a more complex system, involving more information. The thermodynamic cost of keeping such a system running would be relatively lower.

STABILITY

The concept of stability is often involved in discussions of our subject. As implied in the model of expressions (1), the changes in any system through time should be in the direction of increasing constancy in numbers. This constancy may be called stability. It is stability that is the consequence of inter-actions inside the system. But stability as it is commonly used in ecology has something to do with exogenous variations of the parameters, such as E, N_1, N_2, . . ., through agents out-side the system. If such parameters change often (climatic changes, destruction of organisms by unforeseen agents, and so on), then the term stability is frequently used to designate the ability of the system to remain reasonably similar to itself in spite of such changes. Here belongs MacArthur's concept of stability (MacArthur 1955), which presupposes the existence of alternative pathways for energy flow, chosen according to the imposition of external circumstances. These concepts of stability are quite different, and ecologists vacillate between them or are confused by them. To avoid confusion, without complicating the ecological vocabulary it is possible to say directly that in the first case the system is achieving a steady state under constant conditions; in the second case, however, the system has a greater resistance to changes that are external to the system in their origin. The latter resistance is paid for through a higher energy flow and a small number of interacting elements, rather unspecialized in their reciprocal interactions, as shown by uniformly high absolute values of the few significant coefficients in matrix (3), in which the majority of coefficients are zero.

Let us put this again in the language of information theory. In the first kind of stability, the next state of the systems is predictable from within the system, the system contains much information, and new events add small amounts of information. In the second case, the future is less predictable, the system contains less information, and each event represents a relatively important source of information. It is important to remember that a price must be paid for the acquisition of new information. Thus stability in the face of external fluctuation is more expensive than steady state stability.

But, after all, perhaps it is not necessary to use the word 'stability'. Leigh (1965) speaks only of the 'frequency of fluctuations' as a good operative definition of what he has in mind. It is important that the frequency of fluctuations is related to a term in which the ratio production/biomass stands prominently as a factor. The 'frequency of fluctuations' is higher when this ratio is high so that the 'frequency of fluctuations' is related to the inverse of stability in the first (steady state) sense.

DIVISIBILITY OF ECOSYSTEMS

Any model similar to expressions (1) can be subdivided in many possible ways. The distribution of the signs and of the orders of magnitude of the coefficients of interaction is important. The same duality which is recognized in a simple interaction loop can also be recognized in larger assemblages. We can draw a line putting plants on one side and animals on the other, or we can make more subdivisions by applying some concept of the niche. Also, we can separate segments of the ecosystem more realistically by drawing a boundary surface in the real space of the biotope. Across any one of these boundaries, imaginary or real, there is some energy exchange, and the exchanges and properties of the two parts of the eco-system can be averaged so that the two subsystems interact across the boundary in a way similar to a simple feedback loop. There is always some regulation: one subsystem is more con-trolling, the other more controlled, and one of the parts pays a higher energy bill than the other. Any ecosystem under study has to be delimited by arbitrary decision, but one has to re-member always that the imposed boundaries are open, and that the sort of interaction going on across such boundaries is dependent on the respective properties of the two systems on either side of the boundary. With this proviso, all the problems of defining closed ecosystems, limiting superorganisms, etc., happily vanish.

To clarify such ideas, let us suppose that we lump the plants or primary producers in one section, and the animals in the other. It is obvious that such a system can be made equivalent to the previously discussed model of a predator and its prey. The plants supply the energy, foot the entire energy bill, and the functioning between the two sections acts as a gate, allow-ing energy to pass from plants to animals. I feel that it would require too great an enlargement of the concept of niche to apply it to either section of such a partitioning, but it is formally possible. To apply this concept to any subsection joined to others through asymmetrical frontiers seems to be greatly at variance with the usual notion, but I would not refuse to admit the idea.

If we prefer to divide our ecological cake according to real space, we can, for example, separate the benthos of a body of water from the overlying plankton. In this case, although both subsystems obviously contain plants (primary producers) and animals (consumers), there is still some asymmetrical ex-change across the boundary, of the same nature as in the pre-vious case. On the average, the benthos is more driving, con-suming more energy, and the plankton more driven. The elements of interaction between subsystems are as follows: plankton settles down or is attracted by filter feeders, and nutrients regenerated at the bottom and dissolved in the water go back to be used again by the plankton. The result is a net flow of energy in chemical form associated with organic matter toward the benthos. The plankton pays the major part of the bill.

Organization of ecosystems in space is closely linked to the gradients across asymmetric boundaries or to the very existence of such boundaries. It is worthwhile to add some considerations that will help us become aware of the importance of the spatial extension of ecosystems. An ecosystem can be considered as the material path followed by energy that generally enters in the form of photons and goes out into an energy sink. A very simple model of an ecosystem would be formed by some fluorescent substance in which incident photons drive electrons out of their initial orbits. Later the electrons fall back into the lower energy orbits and light of some wavelength is produced. As an example of a real ecosystem with a similarly low degree of organization, we may choose a pelagic ecosystem in turbulent water. The average distance between producers and consumers is small and evenly distributed, and the energy that enters the system by photosynthesis is dissipated in a short time and in a limited space. In other words, the average path of energy-bearing organic molecules is only a few centimeters long, and the cycle of the elements is completed in the same dimensional scale.

In more stable and stratified water, diffusion and mixing are less in the vertical direction than in the horizontal, and the gradient in light and energy input is decisive. The center of gravity of the vertical distribution of animals lies deeper than the center of the distribution of plants, and bacteria are distributed, on the average, still deeper. Here, the average distance between the point at which photons energize some element of the system and the point at which such an element leaves the organisms is of the order of several meters. The longer average trajectories followed by atoms of the different elements in their cycles is reflected in a slowing down of the respective cycles. The ecosystem has grown in space.

An ecosystem, like any persistent structure, has some properties in common with the solid state, expressed in the ability to construct mechanisms. The comparison between a gas and a machine made of solid parts may be helpful. A force applied to a gas transmits movements to its molecules; the small eddies represent a kind of structure, because there is a certain order and some predictability associated with them. But it is a very low degree of organization as far as transportation of energy goes: energy promotes a disorderly movement of molecules and is lost as heat. In a mechanism composed of rigid levers, however, heat appears only at the limits of the solid structures where there is friction, and only here can we find a similarity to the behavior of a gas. But the molecules of a rigid body have limited freedom. They move in an orderly fashion, impelled by energy associated with a structure that does not decay through time. In a mechanism made of rigid parts, energy is transferred between distant points in a thermo-dynamically cheap way.

Now, it is my point that ecosystems behave rather more like solid mechanisms than like gases, in the same sense that eco-systems transmit energy long distances, organize space, and, in doing so, influence future events. As a corollary, a measure of the organization of the ecosystem may be found in the average distance between the place of energy input and the energy sink. The distance can probably be measured either in terms of space or of time; in other words, a system maintaining a very slow cycle over a short distance may be equivalent in terms of organization and of influence on the future with a system running faster but embracing in its cycle a larger space.

Perhaps, in the depths of his psyche, the ecologist may be grateful to cybernetics for throwing a tenuous mantle of respectability over the discredited method of reasoning by analogy. Further steps along the same path now seem justified.

A Basic Principle of Organization

Everywhere in nature we can draw arbitrary surfaces and arbitrarily declare them boundaries separating two subsystems. More often than not it turns out that such boundaries are asymmetric; they separate two subsystems that, although arbitrarily limited, are different in their degrees of organization. There is some energy exchange between the two subsystems in the sense that the less-organized subsystem gives energy to the more-organized, and, in the process of exchange, some information in the less-organized is destroyed and some information is gained by the already more-organized. Probably it is useful from the point of view of general science (but distracting from the point of view of ecology) to remember a few such couplings: gas/Maxwell demon, electrical conductor/semiconductor, atmosphere/sea, environment/thermostat, substrate/enzyme, enzyme/RNA, cytoplasm/nucleus, mesenchym/nervous system, biotope/community, plants/animals, prey/predator, plankton/benthos, agrarian communities/industrial societies.

In all such examples, the second subsystem experiences more predictable changes through time. In so doing it stores information better and is a more efficient information channel. The first subsystem is subject to a stronger energy flow and, in fact, the second system feeds on the surplus of such energy. It is a basic property of nature, from the point of view of cybernetics, that any exchange between two systems of different information content does not result in a partition or equalizing of the information, but increases the difference. The system with more accumulated information becomes still richer from the exchange. Broadly speaking, the same principle is valid for persons and human organizations: any exchange increases to a greater extent the information of the party already better informed. Here it is pertinent to give to the concept of information the meaning of something which is arrived at by a succession of decisions and which influences the future; however, any stricter definition in terms of probabilities would be pertinent also.

The unit in such information exchanges is an event taking place between two elements A and B. Suppose element A has a more indeterminate position and a less predictable future, its position being associated with a more diffuse cloud of probabilities. Such an element carries and can give off more energy. Element B has opposite properties. The interaction event must occur closer to B than to A, and the memory of element A will be annihilated in the interaction. If there is an increase of information content, it will accrue to B. Decision-making means friction, and the bill is footed mostly by A. In this connection, the ideas of Landauer (1961) are very pertinent from the point of view of the ecologist.

Such relations are compounded in a hierarchical organization and are reflected at every level. Any organization can be analyzed by dissecting it into smaller and smaller blocks; the pattern of the whole organization is reflected at every division in differences of organization on either side of the boundary and in the exchange of energy at each level. Such analysis frees us from the need to define ecosystems that are more or less closed.

DIVERSITY

The ecosystem may be considered as a channel which projects information into the future (Margalef 1961*a*). The distribution of individuals into species affords a preliminary measure of the width of the channel of information. One may compare an ecosystem to a message transmitted by means of a certain code. If the number of individuals of species, $a, b, \ldots s$ is $N_a, N_b, \ldots N_s$, and N is the total number of all individuals, the probability that one individual belongs to species i is p_i ($p_i = N_i/N$) and the average information content per individual can be expressed in one of the two following forms:

$$D = -\sum p_i \log_2 p_i, \tag{4}$$

$$D = \frac{1}{N} \log_2 \frac{N!}{N_a! N_b! \ldots N_s!}. \tag{5}$$

Both expressions are borrowed from information theory, but independently of their meaning in that context both are useful as functions with a minimum value, if all individuals belong to the same species, and a maximum value, if each individual belongs to a different species. Both hypothetical limits are equally improbable for a community in nature. The actual values are in between, with a considerable range of variation in what may be called 'diversity'.

It is curious, and almost a perversion of ecological thinking, that the interest of ecologists has been attracted by the regularity in the numbers of individuals of different species present in an ecosystem when enumerated in order of decreasing abundance. A simpler approach to such a series would be to regard it as an expression of the richness and variety of species and, in consequence, of the information-carrying capacity in an assemblage.

A butterfly collector planning a collecting trip might be interested in the diversity of local lepidopteran faunas. An area with greater diversity of butterflies would yield more species, and more valuable ones, because there would be a higher proportion of rare varieties. Of course, this is hardly a demonstration of the utility of the indices of diversity. A research project to map diversity of butterfly faunas aimed at assisting collectors in planning their trips would probably not succeed in attracting financial support. Other viewpoints might be more convincing.

The ecologist sees in any measure of diversity an expression of the possibilities of constructing feedback systems, or any sort of links, in a given assemblage of species. In an ecosystem, a higher diversity means longer food chains, more cases of parasitism, symbiosis, and so on. Nevertheless, diversity is not a complete or adequate expression of the channel width of the ecosystem considered as a transmitter of information. Given the same numerical value of diversity, computed according to expressions (4) or (5), a planktonic population, composed of individuals that continually change their location in space, obviously projects a smaller amount of information into the future than does a benthic or terrestial community in which individuals are fixed in place or move around in a determinate way. On the other hand, the formal analogy of the expressions used to compute an index of diversity from the proportions of individuals falling in different species, with expressions of entropy, does not in itself justify basing thermodynamic properties of the ecosystems on values of diversity indices.

Diversity, as a measure of organization, is a parameter that must be qualified, It is not correct to speak simply of diversity; reference should be made to spectrum of diversity. A spectrum of diversity starts with an individual (diversity = 0). As the sample is progressively enlarged, successive diversities can be computed. Similar spectra can be centered in any point of the ecosystem. Two populations may have similar numerical values of the index of diversity for a given size sample and still have completely different diversity spectra. In some ecosystems diversity increases almost indefinitely as the size of the sample is increased. This means that as the sample is enlarged new species are added and the proportional representation of the different species is always fluctuating; this seems to be the case in the tropical rain forest. Such a situation suggests a hierarchical structure of the ecosystem in space: small groups are organized into larger ones that are all different and these in turn form different and larger structures.

Quite different is the structure of the plankton in turbulent water. Spatial organization is again and again destroyed by turbulence. Diversity may be excessive in small samples but as sample size increases the diversity curve soon flattens and remains nearly constant for a long while. Plant ecologists have dealt at length with statistical problems related to distribution of species in plant communities. As might be imagined, they are much more attracted by communities comparable to the plankton of turbulent water than by communities like the tropical forest. In this connection it must be acknowledged that so far no evidence for 'uniform' or 'random' distribution has been found in natural populations, even in planktonic ones. All points in an ecosystem have unique properties, and it is a bad beginning to assume that one is working with samples from a 'uniform universe'.

Spatial configuration of ecosystems is in itself a very important factor. There are topological limitations to the interaction between species, so that the simple principle of mass action assumed in expressions (1) is not strictly correct. Structure is very important, as is proved in experiments of predator/prey interaction in which spatial patterns of structure are introduced (Huffaker 1958).

DIVERSITY AND ENERGY FLOW

With reference to the whole ecosystem, a useful parameter is the amount of energy flow per unit biomass (the primary production of the system divided by the total biomass). The input energy is nearly all dissipated in respiration. It is also appropriate to speak of the entropy produced in sustaining a unit of biomass in the ecosystem; this entropy is roughly proportional to the total flow of energy. If a system has many trophic levels, the energy flow per unit biomass is lower because a fraction of the energy passes through different levels. In a system subjected to frequent changes in which a high proportion of the primary producer's substance is decomposed by bacteria (heavy users of energy with a small biomass), energy is inefficiently used and relatively more entropy per unit time and unit biomass is produced than in a more diverse and more efficient ecosystem. An equivalent argument may be applied if the ecosystem is considered as an information channel: in changing environments or in poorly organized systems, maintenance of the channel is more expensive because excess fecundity is necessary to make up for the loss of individuals. A high level of noise must be compensated for by redundancy in the message. Such a communication channel is more expensive. It is easy to understand why ecosystems with greater diversity are sustained by a lower energy flow per unit biomass. According to Ashby's law of the requisite variety in cybernetic systems, a system formed by more elements with greater diversity is less subject to fluctuations.

More convincing to the field ecologist is another argument. Imagine an ecosystem for which the opportunities to assimilate and grow are suddenly increased. Not every species responds equally to such opportunities, because the specific rates of potential increase vary among species. Moreover, species dependent on other species—for example, animals dependent on plants for food—respond after a certain time lag to the initial impetus toward the increase of biomass. Because of this fact, any sudden increment in the ratio of production to biomass is accompanied by a drop in diversity. In the opposite case, when nutrients are being consumed while the environment remains constant, the machinery of the ecosystem is acting in accordance with expressions (1) and a maximum diversity is attained. It is to be expected that an increase in biomass, assuming the same primary production, is associated with an increase in diversity. Such changes in diversity can also be deduced from expression (5). It can easily be seen that the substitution of an individual of a common species for an individual of a rare species (a case of a sudden burst of productivity) leads to a decrease in the value of diversity, and vice versa.

If B represents biomass and D diversity, I would like to imagine that both magnitudes could be dynamically related, if only approximately, in the following form:

$$\frac{d^2 B}{dt^2} = -\frac{dD}{dt}. \tag{6}$$

In preceding discussions it has been argued that almost every surface imposed on an ecosystem divides it into two territories with different degrees of organization, one of them with a higher energy flow per unit of biomass. Corresponding to this difference, other differences in diversity are to be expected. Perhaps the gradient of diversity at any point in an ecosystem could be made a function of the gradient of productivity, relating the gradient of diversity (not exactly the same as the spectrum of diversity but related to it), to some expression of the form

$$\frac{d^2 B}{\partial t \, \partial x}, \tag{7}$$

where B is biomass, t is time, and x is a dimension in the direction of the measured gradient. Of course, there would be a flow of the produced biomass from the side in which it is produced at a higher rate toward the side in which the productivity is lower. Let us repeat again that the less-organized system feeds the more-organized.

Such deductive ecology is hardly science and sounds rather

like a scholastic medieval construction. In fact, the motivation for such reasoning has been found in the study of plankton. In chapter three, a number of selected situations and problems which lead to this theoretical model will be discussed. It should not be thought that such intellectual exercise with cybernetic models is useless, for it uncovers some unsatisfactory qualities of concepts in everyday use in ecology. Take, for instance, biomass; a good operational definition may be the amount of living mass, expressed in grams of organic carbon per square meter. But biomass is a bearer of information, and, in cybernetic models, it may be convenient roughly to equate biomass with amount of information. It is also proper in the field of cybernetics to speak of the amount of information preserved and communicated in terms of a given increase in entropy or a given consumption of energy, rather than to speak of amount of biomass. In fact, in ecosystems there are many things besides biomass which are very important in preserving information—like burrows, trails, and so on. Take, for example, the dead wood in a forest, both standing and fallen. It is no longer living matter, but it is tremendously important as a structural element. One might ask whether such wood should be included in the revised concept of 'biomass'.

MICROSCOPIC AND MACROSCOPIC ASPECTS OF THE ECOSYSTEM

Almost inadvertently we have been shifting from the consideration of elementary relations in the ecosystem, like the response of a population to an environmental change or the interaction between a predator and its prey, to elementary cybernetic feedback loops and to the multiplicity and organization of a great number of such feedback loops. Almost everyone will agree that it would be difficult, but theoretically feasible, to write down the interactions between two species, or possibly three, according to the equations suggested by

Volterra and Lotka. This can be done in the ordinary differential form as in expressions (1), or in the more fashionable cybernetic form. But it seems a hopeless task to deal with actual systems; first because they are much too complex and second because we need to know many parameters that are unknown. Thus, we cannot use the magic formula "feed the data to a computer," the panacea of those who expect miracles from the feedback between man and machine. From the practical point of view of the average ecologist, it makes almost no sense to try to write down a system of equations similar to (1) for all the inhabitants of a beech forest.

It is possible, however, to approach the problem from another angle. Biology knows, or pretends to know, something about the organization of the brain; nevertheless, it still cannot afford, and is not particularly eager, to get a detailed description of the net of actual neural connections. Many general properties of the brain can be inferred from the total number of neurons, topological relations, and hierarchical patterns. The number of possible connections and functional levels, etc., can be estimated and are useful in evaluating experimental results. The same is true of ecosystems. It is possible to escape the 'microscopic' analysis of the system, the detailed knowledge of all elementary actions and reactions. A 'macroscopic' point of view is more akin to thermodynamics and cybernetics and permits manipulation of such concepts as energy flow per unit of preserved information, spectra of diversity, and so on. Volterra (1926) hinted at such possibilities. In two very stimulating papers following notions of statistical mechanics, Kerner (1957, 1959) has made considerable progress by introducing a concept of 'temperature'. This is a typical example of a macroscopic property—it is positively related to energy flow per unit biomass and inversely related to diversity.

Besides envisaging a more extensive use of such 'macroscopic' properties or parameters in the ecology of tomorrow, I hope that the consideration of the almost unlimited divisibility of systems and of the asymmetry of the interactions between subsets may be the key to understanding the organization of ecosystems.

2

Ecological Succession and Exploitation by Man

SUCCESSION

Ecosystems reflect the physical environment in which they have developed, and ecologists reflect the properties of the ecosystems in which they have grown and matured. All schools of ecology are strongly influenced by a genius loci that goes back to the local landscape. 'Desert' ecologists, working in arid countries where weather fluctuations exert a controlling influence on poorly organized communities, would hardly accept as a suitable basis for ecological theory the points of view put forward in the preceding chapter. One need only go through the books of Andrewartha and Birch (1954) and of Bodenheimer (1958) to be convinced of this. The mosaic-like vegetation of Mediterranean and Alpine countries, subjected to millenia of human interference, has assisted at the birth of the plant sociology school of Zürich-Montpellier, with Braun-Blanquet as the most representative exponent of a clear and careful bookkeeping and filing system. Scandinavia, with a poor flora, has produced ecologists who count every shoot and sprout. It is a pity that the tropical rain forest, the most complete and complex model of an ecosystem, is not a very suitable place for the breeding of ecologists. And it is only natural that the vast spaces and smooth transitions of North America and Russia have suggested a dynamic approach in ecology and the theory of climax. In such areas, the concept of succession, one of the great and fruitful ideas of classical ecology, was best formulated. An important contribution of field ecologists, it was originated by Dureau, A. P. de Candolle, Humboldt, and others a century and a half ago, and was developed primarily in America at the turn of the century by Cowles and Clements.

Succession is viewed as the occupation of an area by organisms involved in an incessant process of action and reaction which in time results in changes in both the environment and the community, both undergoing continuous reciprocal influence and adjustment. The important point is that it is an orderly and directed change. Unmistakable trends that permit prediction can be recognized, of course, only at the 'macroscopic' level.

In ecology, succession occupies a place similar to that of evolution in general biology. The parallel extends as well to some conceptual difficulties. Everyone seems to agree that evolution is 'progressive'. But definitions of progress are often imprecise or contradictory. The comparison of a series of species that probably represent links of a phylogenetic chain do not always allow one to abstract trends of general applicability. Some paleontologists, aware of obvious directions in evolution, have preferred to gloss over the problem by inventing names like 'orthogenesis'. In doing this, they behave like some ecologists, among whom the belief was, and is still widespread that a good way to conceal ignorance is to invent some beautiful name with a Greek sound. In general, ecologists agree that succession follows a direction, but opinions differ when a working definition, or even some criteria or common characters of the usual trend, are to be formulated in a concrete way.

There exists an important body of data and speculation about succession. The raw data consist either of actually observed successions, including the description of different stages through time, or, more frequently, of an ideal reconstruction based on ecosystems observed in different places at the same time. These 'snapshots' are serialized to produce what is supposed to represent a natural succession. The student of evolution who postulates certain relations of descent among certain species that he knows only on the basis of fossil material is subject to the same sort of criticism as the ecologist who draws an ideal schema of the time succession linking a number of communities that he has observed in different places at one time. In both cases, the proposed orderings are not based on observed sequences, but on abstract criteria that define what should be the expected changes during evolution or during succession. A number of working principles are used to classify and serialize the stages of hypothetical successions. They are not always explicit.

Prominent among the criteria used are the following: during succession there is a trend toward increase in biomass, stratification, complexity, and diversity. Sometimes, too, a maximization of the total production is assumed. Another frequently used criterion is increased stability, that is, a reduced frequency of fluctuation and a major independence from environmental change. In many instances, succession is linked to geographical patterns, and then the importance of local climates for the different stages becomes evident: some climates cause a total arrest of succession such that the community is held in certain stage; under less restrictive conditions, progress may simply be slowed, and the development of the ecosystem itself may help to alter the climate and introduce future development of successional stages.

A New Look at Succession

I have sketched a more or less classical introduction to the problem of succession. Now, if we look at succession from the point of view set down in the preceding chapter, it appears to be a process of self-organization occurring in every cybernetic system with the properties of an ecosystem. A self-organizing system can pass through different states; any change that leads to a state more resistant to further alteration is immediately assimilated. It is probably justified to say that any system formed by reproducing and interacting organisms must go on to develop a kind of assemblage in which the production of entropy per unit of preserved and transmitted information is at a minimum. The structures that endure through time are those most able to influence the future with the least expense of energy. The process of succession is equivalent to a process of accumulating information. The initial, poorly organized, stages receive the full impact of the environment and any changes in it. Individuals of different species are selectively destroyed. The process of acquiring information must be fed by a surplus production of new organisms. Relative energy flow is high and represents the cost of accumulating information. In time the acquired information is expressed in a new organization of the ecosystem. This organization takes into account the predictable changes in the environment, and even controls the environment, so that in the future much smaller changes in the community are necessary to keep it in stable occupation of its area. More information is being transmitted through time, and the environment, as a source of new information, is less important than at the beginning of the process. One can say that the ecosystem has 'learned' the changes in the environment, so that before change takes place, the ecosystem is prepared for it, as it happens with yearly rhythms. Thus, the impact of the change, and the new information introduced, are much less.

It is a general property of many systems that acquired information is subsequently used to close the door to a further inflow of information. In general, the development of a personality involves the use of information to make oneself impermeable to new sources of information; this is, of course, regrettable, especially in the scientific personality. Everyone knows how difficult it often is to make new ideas acceptable to a hard-boiled scientist; only youth is plainly receptive to the many suggestions that come from the outside world. Patten (1961), in a very creative paper, calls our attention to the pertinency of a von Neumann's game, in which one of the two players seeks to gain information from the other to use in blocking his gain of information from the other. Patten adds that the solution of such paradoxes is a fundamental problem of regulation. Translated to the field of ecology, the paradox means that the community seeks to gain and does gain information from the environment, only to use such accumulated information to block any further assimilation of information. This process is succession.

A theory of succession based on such considerations should pass the test of being explanatory and predictive. But prediction has to be limited to the 'macroscopic' level. Prediction in detail is impossible. An ecosystem is a historical construction, so complex that any actual state has a negligible a priori probability. The same is true of evolutionary theory: some general trends can be assessed and future developments can even be predicted, but no one expects a prediction extending to the minor details of structure and function.

CHANGES OCCURRING DURING SUCCESSION

It is difficult to summarize all the available information on succession. I have done this for some marine communities (Margalef 1962b), and many textbooks on ecology offer valuable chapters on plant succession. Some common trends can be abstracted from the data. Of course, such trends refer exclusively to the characters or properties that have been called 'macroscopic'.

Biomass increases during succession as, almost always, does

primary production; however, the ratio of primary production to total biomass drops. Diversity very often increases. Sometimes diversity increases to a certain value and then decreases again toward the final stage of succession. But such situations can probably be better described in terms of diversity spectra. A spectrum with a plateau is replaced by a steep spectrum. Kershaw (1963), Pielou (1966), and other workers have dealt with such problems. According to Pielou, during succession species diversity decreases and pattern diversity increases; that is, diversity and organization shift to higher levels. We might compare a well-developed and mature ecosystem to a picture by an old master—the pattern of a developing system would be rather like some form of op art. (Incidentally, diversity and information theory are highly useful in the analysis of artistic creations.)

During succession there is an increase in the proportion of inert or even dead matter with a low respiratory rate, such as wood, shells, and so on. The proportion of structures like burrows, paths, and territory markers, which may be considered as stores of information, also increases. Fluctuations are damped and rhythms change from reactions directly induced by external agents to indirect responses to stimuli associated with ecologically significant factors; the ultimate trend is to endogenous rhythms.

The increase in diversity is related to a multiplication of ecological niches, a process that goes with longer food chains and much more strict specialization. Judging from data on the feeding efficiency of animals, it seems that animals at the top of long food chains and animals of more specialized habits show a higher efficiency. This results in a major, overall gain in efficiency in the advanced stages of succession.

In later stages of succession, a relative constancy of numbers is achieved, and populations are not forced to reconstruct themselves rapidly after drastic and extensive destruction.

The natural trend is toward a reduction in the number of off-spring produced and better protection for the young. Means of dispersal are also different. Pioneer species are more adjusted to indiscriminate dispersal. In initial plant communities in the succession, a major proportion of the diaspores are carried away by wind, but in later stages plants with seeds dispersed by animals are more numerous. In fact, such plants are geared to animals belonging to the same ecosystem, so that the whole ecosystem develops an organized dispersal. Only in the initial stages of ecosystems is there an important and fluctuating store of precious nutrients in the environment. In the more advanced stages organisms exert more vigorous control, and the major proportion of biogenetic elements are stored or retained by the living organisms.

Ecosystems are composed of replicable prefabricated pieces—individuals of different species. The supply of species is limited; thus, succession cannot go on forever. Succession is an asymptotic process. Only evolution in the framework of the ecosystem can improve things a bit and allow further progress.

THE FINAL STAGES OF SUCCESSION

In ecology it is customary to speak of the climax, with reference to the final stage of a succession, when the ecosystem is in equilibrium with the existing supply of species and the properties of the local environment. The concept of climax has led to controversy over how far changes themselves are to be included in the characterization of environment. This problem is more acute in cases in which populations and ecosystems are processes rather than 'stable' organizations, as in running water. Many ecologists, myself included, are disinclined to speak of climax at all in reference to plankton in rivers. The problem is not serious if we are prepared to drop completely the world 'climax' and speak instead of less and more mature ecosystems. We are then free to define the conditions of maximum maturity to which someone else might affix the label 'climax'.

Succession is not necessarily a continuous process. In the coastal and shallow waters of tropical seas, extensive meadows of sea grasses (*Thalassia* and others) cover large areas. An ecosystem of different structure, the mangrove, then develops on the edge of the water, encroaching on the submerged community of sea grasses. This substitution is another step in ecological succession, but it is linked with the use of a substantially different source of carbon dioxide. In the first stage the carbon dioxide is dissolved in the water in the form of dissociated carbonic acid, but it is in gaseous form in the subsequent stage. Such important and somewhat discontinuous steps in succession have been called 'relays' (Dansereau 1954), a concept that may be of utility.

In terms of the expressions (1) of the preceding chapter, succession is the result of an indefinite iterative process with a stationary end state to which the variables tend. A set of conditions controls a set of rates of change of the same conditions, and the whole system is directed toward an asymptotic steady state. This could properly be called a climax. But unpredictable fluctuations—unpredictable for the organization of the ecosystem at the moment of their onset—can change the values of environmental factors (E), or the numbers of individuals of some species (N_i), altering the smooth running of the model. The process of self-organization stops when fluctuations are unpredictable or insurmountable. In an arctic environment, the hard conditions of life that exist during part of the year limit the accretion of organization. The amount of organization that can be transmitted to the next vegetative period is rather small. This is characteristic of the tundra. Farther south, in the belt of deciduous forests, development of vegetation can achieve a more complex structure that includes trees, but there must still be a period of dormancy and numerous forms of life are excluded. Only farther south, in a part of the tropics, is the environment stable enough to allow succession to go on indefinitely and to achieve ecosystems of a very high level of organization. Some authors, like Connell and Orias (1964), do not agree that the tropics are closer to equilibrium and that the temperate zones are in a successional state of development; their opinion is probably related to their special interpretation of the concept of climax.

Today the most mature systems on earth are found in regions of high temperature: the coral reef in the sea and the tropical rain forest on land. Nevertheless, the point is that stability is a more important factor than temperature, as can be observed by studying the actual distribution of coral reefs. In addition, very mature systems do exist in rather cold environments, provided that periodic frost does not destroy organisms. The deep sea communities, with high diversity (Hessler and Sanders 1966), have all the properties of very mature ecosystems: a low metabolism, limited production of offspring associated with advanced care of the brood, and so on. The communities of organisms living in caves are also very mature. Of course, both ecosystems are dependent for their energy on others in illuminated places, but such dependence is not material to the consideration of those properties that testify to a long process of succession (and evolution).

It is fascinating to contemplate the possibilities inherent in stable environments existing in rather cold temperatures. Perhaps such a situation was achieved in some early period of the history of our planet; perhaps exobiology will some day afford appropriate examples. In a system at high temperature, a higher amount of entropy is produced in the exchange of a fixed amount of energy. In consequence, a system at high temperature cannot go as far (cannot reach the same degree of maturity) as an equivalent system at a lower temperature. Anything that accelerates change and energy flow in an ecosystem causes a reduction in potential maturity. Nevertheless, it happens that at the present time the most stable environments are found in the tropics.

ECOSYSTEMS AS PROCESSES

The study of succession in plankton poses challenging problems. To begin with, the changes observed in a fixed place (sequences) are a combination of true succession and translation (the movement of water masses). If the water were to move uniformly, the problem would be one of choosing between two sets of coordinates, one geographically fixed and the other moving with the water. But a solution is imposed: we must choose a fixed system of coordinates because water movement is too complicated to adjust the study to a set of deformable reference coordinates.

In running water, the same problem is present, perhaps even more dramatically. There are two subsystems in a river —the one suspended in the water and the one linked to the bottom. As the water flows, organisms in suspension are subject to a succession, and more mature stages are found downstream. At a fixed point, however, the ecosystem maintains an appearance of stability of composition. Bottom communities, following the course of the river, can be considered as a succession of stages, since communities upstream influence those in the lower reaches, either by a pressure of colonization (sending diaspores) or by conditioning the water. At the same time, there is no, or very small, influence in the other direction.

Marine plankton, river plankton, and populations in between (those of estuaries), are paradigms of a kind of ecosystem that can be duplicated experimentally in the laboratory in flow cultures or chemostats. It is out of the question to envisage a moving system of reference coordinates. Consider a river, in which suspended organisms multiply. The water flows, carrying away organisms. If the flow were perfectly laminar everything would be washed away and the water would become empty of life. But flow is turbulent, and some organisms actually move against the main current; others are carried away at a speed higher than the mean, and all, of course, multiply. As a result, at a geographically fixed point a population is maintained in which increase by multiplication compensates losses by drift, diffusion, and sinking. The population has to be considered a process rather than a state or organization; it is like a cloud that forms at one end and disappears at the other, maintaining in between a form and a certain appearance of organization. In dealing with such a system we are led to propose a very broad approach, referring events to a system of geographically fixed coordinates and keeping every element of the ecosystem open, with fluxes and exchanges across all the limiting surfaces. Different equilibria between local accelerations and decelerations of the different processes may add up to changes that can be described in terms of maturity.

Suppose we have a uniform suspension of several organisms, and a number of them move across a division line from one half of the space to the other half. The population experiences dilution in one part and an increase in concentration in the other. This experiment can be easily done in a U-shaped culture vessel in which the branches are separated by a fritted glass filter that allows passage of the fluid but not the cells. If, after a reasonable time-allowance for development of the mixed population, pressure is used to pump some water across the filter, increasing the volume of one side at the expense of the volume of the other, we have achieved an expansion and a contraction of equivalent systems. I have often done this experiment in practical work with students, and the more dilute subsystem always reacts rapidly and shows characteristics of diminished maturity (lower diversity and a higher ratio of primary production of biomass). Simultaneously, the maturity of other population increases. It is easy to understand that the rate of division is reduced among the more crowded cells and that the more dispersed cells are free to multiply faster; in doing so, the latter show an increase limited, at least in the beginning, to one or a few dominant species. The experiment is, of course, trivial, but it suggests further work in the study of the relations between organization, space, and movements.

EXPLOITATION

One important contribution of the study of plankton to general ecological theory arises from one of the most essential properties of pelagic ecosystems, their subjection to steady exploitation. The plankton in superficial water is an expanding community; a fraction of planktonic organisms is continually settling down. This continuous leakage of a part of the biomass requires a higher rate of primary production per unit of remaining biomass, and selection pressure favors a few species with a high rate of potential increase, so that diversity is rather low. Expressions (6) and (7) of chapter one are applicable here. Plankton in surface water, like plankton in running water, is in a state of low maturity; for this reason, I am reluctant to speak of climax in reference to such systems.

Plankton in surface water is exploited by communities in deeper water (the benthos), and populations in the upper reaches of a river are exploited by populations downstream. Now we are able to link the concept of maturity with the concept of exploitation. A more mature system always exploits a less mature system, and the notion of exchange across a given boundary may be helpful in assigning the adjoining systems to their proper places in an ideal succession. The problem can be put into an appropriate perspective by the following reasoning. Suppose we have two adjacent systems of different degrees of maturity. If succession proceeds, any line marking a particular value of the 'macroscopic' characters used as indicators of maturity (diversity, ratio of primary production to biomass) should move toward the system that was and is of lower maturity, since the maturity of the whole complex of subsystems is increasing. But if there is a strong exploitation of the less mature subsystem by the more mature one, the line may not necessarily move. This is because the excess production, which the less mature ecosystem could use to increase its own maturity, is being transmitted to the other subsystem. Thus, the less mature subsystem is kept in a steady state of low maturity by the exploitation to which it is subjected. In nature things are probably not so clear-cut; there is always some exploitation going on, but part of the surplus is used in increasing local maturity.

In future development of a quantitative treatment of this problem, we must try to unite both autonomous (passive) movement of organisms and active exploitation on the part of the other subsystem in any formula expressing exchange across a boundary. Moreover, we must include exchanges of nutrients and organic matter outside the bodies of living organisms.

Successions are stopped by fluctuations in the environment, which may be equated to exploitation. After all, a frost that destroys living tissues, or natural catastrophes that do away with a high fraction of the population, really represent exploitation. Exploitation has a rejuvenating effect on exploited ecosystems. It is sufficient to add a protozoon to a senescent culture of algae to observe a rejuvenation in the plant population, with a higher energy flow per unit biomass and a more juvenile composition of plant pigments. In a way, the addition of a new trophic level to a system means a certain amount of rejuvenation of other trophic levels, increasing turnover.

Plankton is the ecosystem in which the tenuous equilibrium between succession and exploitation can be best observed. Changes in the turbulence and stratification of water masses switch the process one or the other way. In turbulent water there is a large loss of plant cells, and the system is juvenile. If water stratifies, succession proceeds immediately but slowly, because there is always some degree of exploitation. Only in stratified water masses are relatively mature plankton systems found.

BOUNDARIES

Descriptive ecology uses maps with patches of color intended to represent different kinds of communities or ecosystems. Ecology is, or at least has been, very prone to formalism, and a word, 'ecotone', was introduced by Clements to designate the ambiguous boundary between patches. Happily enough, in the etymology of the word there is a reference to tension—a suggestion of something dynamic that can give life to the patches of a map. Indeed, such boundaries must be considered places of tension where two organizations meet and exchange their respective components, or as places where stresses of a genetic character, important in evolution, are at work. In practice, however, ecotones often disappear when we look for them. Descriptive ecologists often base descriptions of communities on typical spots observed in the field, tracing in the boundaries as homework. In fact, we do not really need marked frontiers. It is enough to observe that the composition of an ecosystem is different in two spots to assume that some change must occur between them—perhaps gradually, perhaps more steeply.

Not all boundaries are the same. Some may be drawn between patches of different specific composition but of equal maturity. The most important are those separating sub-systems of different maturity, like the ones delimiting concentric belts of vegetation around a senescent lake or those running parallel between a sand dune and the forested land behind it. Although we were disposed to save it, the expression 'ecotone' is denied to us in this case by Shelford, who expressly declares in his *The Ecology of North America* (1963) that he will not use or apply the term 'ecotone' to the boundaries between communities that represent different stages of a succession or serial transitional communities. Sometimes ecotones that are very clear, like those along shorelines, separate ecosystems that are different in many respects, including degree of maturity. Here it is pertinent to remember that if two ecosystems in contact are of very different composition, transfer across the ecotone is in the form of detritus. This is how exchange is established between marine and terrestrial ecosystems along a shore or between marsh vegetation and the aquatic organisms living in estuaries.

In a preceding section, the relative importance of exchange in relation to the mobility of ideal or visible boundaries has been discussed. Now some attention should be given to the detailed outline of the boundary. According to van Leeuwen (1965), more undulating boundaries are found in more mature systems and less complicated frontiers in the less mature. But what happens if a more mature system faces a less mature one? The forces at work on the outline of the boundary should depend on general properties of the respective systems: higher mobility, especially random or diffusion mobility associated with a more rapid turnover in the less mature subsystem; more rigidity or determinism in position and organization in the more mature. Probably some concept analogous to surface tension could be developed and applied to the characterization of ecosystems. I suggest that such analogs of surface tension would decrease during succession.

It seems that it is advantageous for the more mature and exploiting system to develop sinuosities and stretch the length of the boundary to the maximum. Just as intestinal villi absorb the contents of the intestine, some benthic populations develop digitations that make the exploitation of suspended systems easy. In the opposite case, in less mature communities the trend must be to reduce to the minimum the extension of the boundaries that are potential sites of exploitation. These conclusions were suggested by consideration of a pattern of distribution observed again and again in planktonic ecosystems. If there are patches of different maturity, they are not disposed in a mosaic but rather follow a honeycomb pattern, with precisely discontinuous areas of lower diversity enclosed in a honeycomb of higher diversity. I have wondered about the forces in action in this case when flying over an expanse of clouds, where similar patterns can be observed. Systems of circulating cells, with convergences and divergences, may appear through action of wind, and even, on a very small scale, by the activity of organisms themselves (Platt 1961, Crisp 1962). In such a case there is a tendency to form a honeycomb pattern, and the dispersed phase with expanding populations is represented by the divergences.

FRESHWATER LAKES

The traditional classification of freshwater lakes arranges them in a series of increasing 'eutrophy', whatever this term may mean. In most limnological writing it is assumed that there is a natural succession from oligotrophic to eutrophic lakes. Such succession is accelerated by human influence. The ratio of primary production to biomass is higher in eutrophic lakes, and their species diversity is lower (Margalef 1964). Plankton of eutrophic lakes contributes relatively more to the sediment; that is, it is relatively more exploited. Nevertheless, absolute production and absolute biomass are usually higher in eutropic lakes, and such parameters are part of the definition of eutrophy.

According to the theory developed here, oligotrophy should succeed eutrophy, not precede it. But the conflict is only apparent. Oligotrophic lakes are certainly more mature ecosystems than eutrophic ones. Eutrophic lakes are kept in a state of low maturity because nutrients are continuously forced into them—by runoff from fertilized crop land, urban pollution, and other human actions—without allowing them enough time to become equilibrated. On the contrary, the inflow of nutrients is accelerated. A decisive experiment would be to cut down the input of nutrients to a eutrophic lake and follow its evolution. I am sure it would change in the direction of oligotrophy. The problem is complicated by another factor, the filling of lake basins, which reduces the hypolimnion and, at a given moment, may represent a relay. In freshwater environments, self-destruction may work against self-organization.

In fact, lakes cannot be studied independently but must be associated with other peripheral ecosystems. More apparent difficulties lurk here. How is the exchange between a lake and its surroundings to be considered? If a lake receives nutrients from agricultural land around it, is it possible to say that the lake exploits the surrounding terrestrial ecosystems? In this case, to be consistent, we would assign a higher maturity to the lake. Such questions may be helpful in clarifying the notion of exploitation. The exploitation of a less mature by a more mature system always concerns organisms, alive and perhaps dead, but not inorganic or even organic and dissolved nutrients. Deep water communities receive organisms and detritus from more superficial communities, and these receive dissolved nutrients from deep water. The cycle is closed, but we assume that populations of the benthos and deep water plankton exploit surface plankton, and not the contrary. To be consistent we cannot say that a eutrophic lake is exploiting the surrounding country. But in certain cases a lake receives a large amount of material in form of detritus; such is the case in certain dystrophic lakes. The incoming material may help sustain a large animal population, and the ratio of primary production to total biomass may be low. Thus, dystrophic lakes may be considered more mature than oligotrophic ones.

ECOSYSTEMS ON A SOLID SUBSTRATE

The theory of succession has been developed mainly with reference to terrestrial ecosystems; in consequence, there should be fewer problems of application than in the case of pelagic ecosystems. But a few questions are really challenging and their examination may contribute to the general theory.

Benthic communities have some similarities to terrestrial ecosystems. The benthos on a soft, shifting bottom is less stable, has less possibilities for constructing organization, and must be considered less mature than benthos fixed on rocks. The diversity of the epifauna on a hard bottom is always higher, and increases from high latitudes to the tropics; diversity of the infauna in soft substrate is generally lower and does not show an appreciable latitudinal gradient (Thorson 1957). According to our general model, diversity of the communities seems to be negatively correlated with turnover.

In terrestrial ecosystems the transport system in the body of plants and the mobility of animals allows us to assign a dimension to the spatial organization of the ecosystem. During succession, average distance between the places of energy input and the energy sink increases, just as in plankton. Compared to aquatic ecosystems, terrestrial ones are notable for an excessively low ratio of animal biomass to plant biomass. This is puzzling because this ratio increases during succession and is rather high in mature systems; we usually consider terrestrial ecosystems as highly mature and organized, in comparison to aquatic ones. Has it always been so? Was the ratio of animal biomass to plant biomass as low in mesozoic times as it is now? There seems to be some historical accident involved. Evolution of animals has not been very successful in utilizing the huge development of structural glucids in plants. In fact, the utilization of lignin and cellulose by animals is largely mediated by symbionts, or involves passage through a detritic form, a still more wasteful process.

The highest ratio of animal biomass to plant biomass is not found in the forest but in grasslands or parkland in which vegetation is kept rejuvenated and exploitable by grazing, as in the prairie of America and some areas of Africa where the biomass of large mammals may range from 3.4 to 18 g per square meter (Clark and Bourlière 1963). In such examples, the most mature *vegetation* is obviously the forest, but we cannot tell whether the most mature *ecosystem*, as a whole, is the forest or the area sustaining the maximum animals biomass. Interesting theoretical problems arise if, as a criterion of maturity, we use the ratio of information preserved per unit energy flow instead of the amount of biomass preserved per unit energy flow. What are the relative contents of information per unit biomass in plants and in animals?

Applied Ecology

In some fields of natural science, the pure scientist sometimes looks disdainfully at human artifacts or at the products of cooperation between man and nature. Asking a serious botanist the name of a flower can elicit the scornful answer, "this is only one of those wretched garden plants!" They are simply not interested. The ecologist cannot adopt such an attitude, but this is not to be construed as a special open-mindedness toward practical problems. Ecologists, like other scientists, are mainly motivated outside the practical sphere. But ecologists must be interested in practical problems for truly scientific reasons.

It is rather difficult to understand the workings of a very complex system like an ecosystem through a detailed input–output analysis; a good study procedure is to subject the whole system to more or less important changes of a very gross nature, to see if and how the general behavior is changed. In other words, the black box of the system has to be shaken. Such a study would be extremely expensive for the ecologist to carry out in natural systems, but he can profit from the drastic changes that man is inflicting every day on all sorts of ecosystems around the world, try to understand their reactions, and use such valuable information for the construction of ecological theory. Besides, as a justification to agencies sponsoring research, ecologists can claim, mostly legitimately, that their findings are invaluable for practical questions of exploitation and conservation.

HUMAN EXPLOITATION

The most direct action of man on nature is exploitation. Man can be considered as a subsystem coextensive with the exploited system. Exploitation reduces maturity or puts a brake on succession. In the exploited system diversity drops and the ratio of primary production to biomass increases. Take, for instance, agriculture. Exploitation of crops means a simplification of the ecosystem in comparison to preagricultural stages. The exploited ecosystem is composed of a lower number of species and also a lower number of biological types (grasses, forbs, trees, etc.). The structure of soil is simplified and the diversity of populations of soil microorganisms and animals decreases. The importance of the cycle of nutrients outside the bodies of organisms is amplified. Yearly rhythms become more stressed, not only for the cultivated species but also for species associated with culture as weeds or pests. The latter are biologically more similar to cultivated species than to species in natural ecosystems. They increase rapidly in numbers, disperse easily, and their populations are subjected to strong fluctuations and can be reconstructed after heavy losses. Outbreaks are characteristic of systems with low species diversity (Pimentel 1961). If exploitation and culture are discontinued, succession starts the reconstruction of more mature ecosystems.

The basic consequences of exploitation and culture can be duplicated nicely in experiments with aquatic systems, most appropriately in introductory ecology courses. Two aquaria, freshwater and marine respectively, are installed and inoculated with material of different origin. After a few weeks or months a relatively well-balanced community is found in them. Species composition, diversity, plant pigments, primary production, and total biomass are recorded. Both aquaria are then subjected to the analog of plowing a field and fertilizing it. Small and equal amounts of a solution of nutrients are poured into the aquaria, and their contents vigorously stirred. The whole set of analyses is repeated the next day. A few species will increase rapidly, diversity will go down, chlorophyll-a will be synthetized at a higher speed than other pigments, and the ratio of primary production to biomass will increase dramatically. These results are a good analogy of what happens in a field subject to culture. It is a good point to make students notice by themselves that the changes are the same in both aquaria, although they have no species in common. This leads to useful discussion of 'microscopic' versus 'macroscopic' parameters in ecology. In agriculture too there are microscopic aspects, such as those emerging from the study of the effects of thermal and rain cycle on plant development, the use of fertilizers, and so on. But the purpose of the experiment is to emphasize macroscopic parameters as well as to demonstrate their importance to practical and applied questions.

Human exploitation in a sustained form is possible only if carried on in an ecosystem of low maturity, with a relatively simple structure and a high ratio of primary production to total biomass. Productivity, or primary production per unit biomass, is maximum with almost no biomass, and decreases steadily up to the maximum biomass. Maximum total yield is possible at some point in between, where productivity multiplied by biomass is maximized. Ecosystems subjected to natural fluctuations can resist human exploitation without collapsing and without much change. The taiga is much more exploitable than tropical forest, and populations of clupeoids (sardine, herring) are much more exploitable than populations of fishes of the coral reefs.

When subjected to exploitation, systems of higher maturity regress and acquire properties similar to those of some preceding stage. Intentional or unintentional exploitation of forests leads to a simplification of the canopy composition and a relative increase in the participation of the fastest-growing species. In general, exploitation of mature terrestrial ecosystems leads to an open vegetation, similar to that of steppes and grasslands, which is subjected to periodic fires, heavy grazing, or other natural agents of exploitation. Man has substituted himself for fire and large grazers, so human evolution and acculturation have been favored in this sort of environment. Above all, temperate climates with ecosystems of rather low maturity have proved especially able to sustain exploitation. Moreover, a large store of wood and cellulose awaited man because it had been poorly exploited by animals. Without doubt, general ecology has something to say concerning the evolution of humanity.

CONSERVATION

A strong exploitation of very mature ecosystems, like tropical forests or coral reefs, may produce a total collapse of a rich organization. In such stable biotopes, nature is not prepared for a step backward. Man has to be very careful in dealing with systems of high maturity.

The interference of man goes farther than simple exploitation. Even if man introduces into natural ecosystems nutrients or a potential source of energy in the form of organic compounds, the effects are similar to those of exploitation. There is an acceleration of energy flow and a simplification of structure, with destruction of many homeostatic mechanisms. Eutrophication of lakes is a good example. Radioactivity destroys structure but does not greatly affect energy flow; the ratio of primary production to biomass is increased and the system becomes less mature (Platt 1965) as was predicted (Margalef 1963*b*).

The species that suffer most from human interference are those endowed with a low rate of multiplication and linked by multiple, if sometimes tenuous, bonds to other elements of the ecosystem. Such species often provide interesting examples of mimicry and other elaborate defenses and impress conservationists by their 'exotic' appearance. The use of pesticides, mitotic poisons, and radioactivity affects such species first, before species competing more directly with man in the use of primary production.

Any human intervention in nature, even presupposing good intentions, can rarely be reconciled with the idea of strict conservation. True, homeostasis in natural systems is always active, and although man tries again and again, he rarely produces truly catastrophic changes in the biosphere. But this is a credit to the efficient organization of ecosystems and not to the wisdom of man. Genuine conservation forbids any interference.

Any project of limited conservation must state an end; sometimes it is simply not to push exploitation farther than a point of adequate yield. In other cases the preservation of the structure of the exploited system above a definite level can be added as a condition. In any case, the study of macroscopic properties of systems, and perhaps just of the spectra of diversity or the quality of pigments, will provide useful criteria for controlling and judging exploitation and the results of conservation measures.

It is impossible in these times to develop a 'natural' ecology, one that ignores the impact of man. Ecology should inspire a wiser management of nature: the feedback should work. Few natural populations have been sampled so extensively as those of commercial fishes, yet some criteria used in fisheries management belong to old ecology. The ecosystem is analogized to a machine—this is the input, that the output. But a new and important notion is that the working of the machine is related to the energy passing through it. Pumping more energy in and out of a system simplifies it; it becomes different and works differently.

Straightforward exploitation of ecosystems has assisted in the selection of their components for productivity. Man has further developed adequate organisms to exploit grass communities and to serve as a link in the food chain ending with man. Efficiency can undoubtedly be increased but proposals like the one to use indigenous phytophagous animals for the exploitation of African ecosystems have probably been motivated by the desire to preserve certain species while appearing to be generous to native human populations. Such proposals merit close attention, although in my opinion they are consistent neither with good exploitation nor with conservation.

To sum up, exploitation as opposed to conservation is a great dilemma, and there is probably no solution that could satisfy the aims of both. Ecology can devise most efficient means of exploitation, but conservation requires non-interference with nature, even refraining from "protecting" her. Probably the best solution would be a balanced mosaic, or rather a honeycomb, of exploited and protected areas. Conservation is also important from the practical point of view: lost genotypes are irretrievable treasures, and natural ecosystems are necessary as references in the study of exploited ecosystems. Moreover, mature ecosystems are factors of stability. The International Biological Programme has at least been useful in pointing out that the conservation of nature has utilitarian aspects and that there may be a stronger motivation for promoting it than the usual esthetic and sentimental reasons.

3

The Study of
Pelagic Ecosystems

INTRODUCTORY REMARKS

In conformity with what was said in the second chapter about the influence of the local environment and ecosystems on the development of ecological thought, I must acknowledge that my own convictions on ecological matters are in part an outgrowth of some observations on marine and freshwater plankton that have impressed me considerably. Any critical evaluation of the ideas discussed and their applicability to structure and function of ecosystems must allow for such an initial bias. And, of course, a more concrete discussion of the evidence should be centered on the pelagic environment. In what follows, a summary of pertinent data is given. Many have been obtained personally and are already published, but the full details have appeared in rather obscure journals and in a language of secondary importance in scientific affairs. They will therefore bear some repetition.

The basic characteristic of the pelagic ecosystem is that it is subject to natural exploitation and to turbulence. Both agents keep maturity low. A system of fixed geographical coordinates is selected for study, although in some respects a system of coordinates moving with their populations would be appropriate. A Lagrangian approach, however, would be much more complicated than a Cartesian one, probably hopelessly so.

51

SPECIES DIVERSITY

The determination of species diversity in pelagic systems is not a simple enterprise. To begin with, it is impossible to have a complete census of the components of a 'cell' of such an ecosystem, from bacteria to fishes and whales. In practice diversities have to be computed in segments of the ecosystem. One soon gets the impression that total diversity is almost mythical, but that diversity of the ecosystem is reflected with little distortion at several levels, so that if the diversity of phytoplankton is high the diversity of zooplankton and even of pelagic fishes is high also. It is not an exaggeration to say that a look at the fish market allows a first estimate of the diversity of the plankton populations living in the same water. A proviso should be added to this affirmation: diversity of the whole ecosystem is reflected in the diversity of its components belonging to a definite taxonomic group only when the spatial distribution of that group cuts across the whole ecosystem. Thus, phytoplankton is a good indicator, but diatoms and dinoflagellates are much too specialized in rather opposite directions for their diversities to be a good indicator of total diversity. Cells are better units to use than colonies, although the tendency of cells to aggregate in colonies is an important factor in the diversity of small samples at the lower end of diversity spectra.

Many samples of plankton are obtained with a net, which exerts a selection on natural populations. For phytoplankton, counts obtained with the help of Utermöhl's microscope are appropriate, but a great fraction of the small organisms in the samples turn out to be impossible to identify by species name. Therefore, exact computation of diversity is hardly feasible for samples of total plankton observed with the inverted microscope.

In Figure 3 some values of diversity calculated from Utermöhl samples are compared with the values of diversity of phytoplankton obtained in the same water with a fine-meshed net. There is a correspondence between these measures of diversity. The reason why so few data are compared is obvious: it is rarely possible to find samples of sedimented plank-

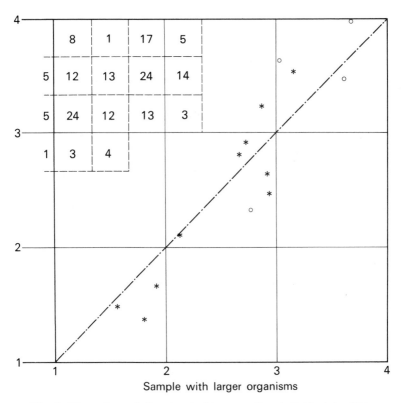

Fig. 3. Comparison of diversity, in bits per cell or individual, in different samples obtained from the same ecosystems. *Asterisks:* Diversity in samples of demersal fishes from the western Mediterranean obtained with a net of 15–20 mm between knots (abscissa) versus diversity in samples obtained with a net of 8 mm between knots (ordinate). *Circles:* diversity in samples of phytoplankton from the western Mediterranean obtained with a net of meshes of 50 μ (abscissa) versus sedimented plankton studied in Utermöhl's microscope (ordinate). Squares in upper left corner; zooplankton of Lake Maggiore, Italy; each square contains the number of cases falling in it. Abscissa: plankton collected with a net of 300 μ mesh size; ordinate: plankton collected with a net of 70 μ mesh size.

ton preserved so well and with such a dominance of recognizable forms that diversities can be computed from the results of the census (Herrera and Margalef 1963). Also summarized in Figure 3 are some data on the correspondence of values of the diversities of freshwater, net-caught zooplankton samples obtained from the same mass of water but with different nets, one of 70 mesh size and another of 300 mesh size. There is a positive correlation, although it is somewhat diminished because one net was more effective on rotifera and the other on crustacea and because both groups behaved nonuniformly in the course of the yearly cycle. In fact, for both sets of data, the different pattern of the distribution of diversities according to area and site gives insight into interesting dynamic properties of the total population (Margalef 1962*a*). In other words, neither the rotifera nor the crustacea are perfect representatives, by themselves, of the whole organization of the ecosystem because they belong to partial 'niches'. For the same reason, neither the diatoms nor the dinoflagellates are very good indicators of the total diversity of phytoplankton (Margalef 1961*b*); in fact, diversity of the partial population of diatoms drops in summer, when diversity of the whole phytoplankton community is at its maximum. Another example of the eventual correlation between diversities of collections extracted from the same ecosystem by different sampling procedures concerns samples of demersal fishes caught with different mesh sizes (Margalef 1965*a*). Results are again plotted in Figure 3.

Ordinarily, the species distribution of individuals is fairly regular. Such regularity allows the use of simplified expressions as indexes of diversity, provided that the number of individuals belonging to the different species, when written in order of decreasing abundance of species, can be approximated by some mathematical expression. Such a measure can be used in place of the more adequate measures of diversity which are based on information theory and in which numbers of individuals of every species present should be entered. A simplified diversity index has the form

$$D' = (S-1)/\log_e N, \tag{8}$$

where S is the number of species and N the number of individuals. There is a good correlation in phytoplankton between D' and the D of equation (5) in bits per cell significant at $P = 0.0015$ and with coefficients ranging from $+0.66$ to $+0.97$ (Margalef 1961b). It is curious that although the correlation is equally significant for diatoms, dinoflagellates, and total phytoplankton, the regressions are not exactly the same for diatoms and dinoflagellates. For the same values of D in bits per cell, the collections formed by the diatoms give lower values of the index D'. It is probable that frequent aggregation of diatom cells in colonies and a stronger dynamism in their populations introduce small differences into the regularities of species distribution of individuals. The correspondence between D and D' fails only in abnormal situations, such as conditions of strong mixing of different water masses in the Mediterranean (Margalef and Herrera 1963). This sort of relationship is important because it shows that normal development of populations leads to a convergence of D and D'; that is, natural populations tend to show a notable regularity in the species distribution of individuals.

Spectra of diversity seem to be related to spectra of turbulence. In turbulent water, diversity in small volumes is low, but is still exaggerated because many species from neighboring spaces are forced together. Their populations are not in equilibrium and, at the same time, the mixing makes the plankton uniform over large areas. Some examples have been worked out in the Ria of Vigo in north-western Spain, although in a rather rudimentary way. A great number of net samples were taken at different stations and diversities were computed for each sample; the diversities were then averaged for all stations occupied simultaneously. The samples collected on the same date were pooled to give an estimate of the representation of plankton in the whole Ria, and an index of diversity computed from this list. Here are some values (not in bits per cell).

Date	Average diversity (D') in every station	Average diversity (D') in the Ria	Difference
May 9, 1955	4.47	7.44	2.97
June 27, 1955	1.53	3.54	2.21
July 30, 1955	3.58	7.01	3.43
August 25, 1955	3.22	5.90	2.62
September 26, 1955	2.90	6.26	3.36
October 24, 1955	2.43	6.28	3.85

According to hydrographical information (Vives and López-Benito 1957), the periods of maximum mixing were the end of September, and the end of October; at these times populations started to multiply and hence had low diversity. To judge the influence of mixing on the change in the spectrum of diversity is more difficult, because different populations remained in small interior bays and it is uncertain how to estimate the significance of differences in diversity indices.

When measured in bits per cell for populations of phytoplankton represented by samples of hundreds or thousands of cells collected by standard oceanographic sampling procedures, diversity fluctuates around 2.5 bits in actively growing coastal populations, is lower in estuaries, and is close to 3.5 and 4 in later stages of succession in more stable water (Margalef 1967*b*). In freshwater lakes diversity ranges between 1 and 2.5 bits per cell in eutrophic lakes and up to 4.5 bits per cell in oligotrophic and dystrophic ones (Margalef 1964). (The papers cited summarize data from other authors.)

The diversity of zooplankton in Lake Maggiore (Margalef 1962, and Fig. 3) ranged from 2.6 to 4 bits per individual for rotifers and larval crustaceans and from 0.8 to 2.2 bits per individual for larger and adult crustaceans. As for the populations of demersal fishes trawled along the Mediterranean Spanish coast, the range is from 1.4 to 3.5 bits per individual (Margalef 1965*a*).

It is noteworthy that all values of diversity fall in a rather narrow range, with an upper limit not far removed from 4.5 bits per individual. Is this a limit of efficiency in the construction of natural homeostats? If such is the case, reasons for this limit should be found in the relations between the possibilities of interaction and the capacities for survival. It is interesting that alphabets also tend to have an asymptotic information content per symbol of approximately the same order of values.

As the study of diversity seems promising, it is worthwhile to explore the use of some equipment that makes such study easier. Plankton could be concentrated and the image projected on the screen of a microscope. A planktologist—still a very important part of the process, but unfortunately an increasingly rare species—can give names to the individuals scanned in a regular sequence corresponding to the sequence of water samples being pumped in. The identification, coded on punched tape or otherwise recorded, could be used as input for a computer programmed to give a spectrum of diversity centered at every point of the explored transect.

Another approach would be to look for a means of evaluating structure which does not require the help of an experienced planktologist. Present-day dimensional particle counters afford a new possibility. The spectrum of distribution by size may be related to actual diversity in the sense that communities with a higher species diversity may have also a higher percentage of big particles. As yet no full data are available for a test, but we have examined the correlation between the percentage of large particles among total particles and the ratio of pigments absorbing light at 430 mμ to those with a peak at 665 mμ (Margalef, Herrera, Steyaert, and Steyaert 1966). This pigment ratio, symbolized as D_{430}/D_{665}, is positively correlated with species diversity. The particle index–pigment index correlation for a series of 127 measurements in the Tyrrhenian Sea was found to be $+0.27$, a highly significant value.

PIGMENTS AND PIGMENT DIVERSITY

The notion of diversity can be applied to anything that can be distributed into categories, including organic matter distributed in different chemical compounds. Pigment diversity seems to be an adequate expression of the complexity of photosynthetic systems. The basic point is that the amounts of different pigments are not independent. Chlorophyll-a has a physiological position that causes its abundance to fluctuate more rapidly than other pigments according to the carrying capacity of the environment. On the other hand, in conditions unfavorable for growth, carotenoids are more resistant to destruction than other pigments. Senescent plants and old cultures of algae turn yellow. The composition of their pigments shows a higher diversity.

Many technical problems complicate efforts to measure pigment diversity. In aquatic ecology it is usual to extract samples with an organic solvent and run optical absorption spectra of the extracts. The expressions used for computing concentrations of a few pigments assumed to be combined in the solution are not satisfactory. Thin layer chromatography proves that the variety of pigments present is higher. I have been using the index D_{430}/D_{665}, representing the ratio of the absorbances at these wavelengths, expressed in millimicrons, of the acetone extracts of the material. The index is correlated with the proportion of yellow pigments and with the general diversity of the pigments. The selection of the wavelength 430 mμ was due to an independent circumstance. It is the band of maximal absorption of Harvey's old standard for the estimation of pigments, and as we used it extensively in our laboratory in converting to spectrophotometric reading, we kept records for a while of optical density at 430 mμ in order to find appropriate conversion factors. The reason for the selection of 665 mμ is obvious; it is close to the peak of chlorophyll-a in the red. Iizuka, Tanaka, and other Japanese workers, probably by the same motivation, proposed an index very similar: D_{435}/D_{670} (Tanaka *et al.* 1961).

As a result of dissatisfaction with present computational procedures, there is some tendency to restrict measurements to the band of maximal absorption of chlorophyll-a in the red and to express the results in terms of chlorophyll-a, leaving out other possible pigments altogether. This idea has even been recommended by international bodies; it amounts, however, to throwing out the baby with the bath water. From any ecological point of view, quality of pigments is as important as quantity of one of the pigments. In fact, the diversity of pigments is an important indicator of the history and activity of the ecosystem. Moreover, to read absorbancy at 430 mμ does not require a tremendous amount of added work in addition to the measurement at the customary 665 mμ. I agree that the concentrations of chlorophyll-b, chlorophyll-c, astacine, and nonastacine carotenes, and even of chlorophyll-a, given in most papers are probably meaningless. But dissatisfaction with present methods should not diminish interest in securing some indication of the quality of the whole pigment complex. As computed by the old methods (from absorbancies at selected wavelengths), concentrations of other pigments relative to concentrations of chlorophyll-a turn out to be correlated primarily with the index D_{430}/D_{665}. Also, if one could suppose that computation of pigments in the traditional way is not totally useless, the simple index D_{430}/D_{665} can do the same service.

In the sea, values of the index D_{430}/D_{665} range from 2.5 to 3.5 for plankton blooms in upwelled water. For freshwater lakes and in similar situations (eutrophic lakes) values can sometimes appear lower if green algae are dominant, but not necessarily so. In older planktonic populations with a greater proportion of dinoflagellates, values of the index up to 8 and more are commonly found. This is most consistent with our approach.

A good positive correlation between species diversity and pigment index has been found. In the western Mediterranean (54 pairs of data), $r = +0.40$; in the Caribbean (68 pairs of data), $r = +0.30$. If phytoplankton is abundant and growing rapidly, the pigment index is low. In a series of 577 pairs of values from the Carribbean, a correlation of -0.34 has been observed between the concentration of chlorophyll-a and the ratio D_{430}/D_{665}. In conditions of high abundance of phytoplankton there is usually dominance of one or a few species and specific diversity is also low. The correspondence between pigment diversity and species diversity should not be interpreted as a consequence of a direct relation between the number of pigments and the number of species, in the sense that more species offer more possibilities of biochemical differences between them. It depends only on similarities of behavior in systems formed by different elements, when these are subjected to accelerations and decelerations in the processes of synthesis and destruction.

The vertical distribution of pigment ratio, as compared with species diversity, ordinarily shows some nonconformities that stress the independence of both characters. At the level of pycnoclines there is often an increase in production with a decrease in both species diversity and pigment ratio.

But sometimes at this level, and more often in deeper water, a decrease in pigment ratio is observed that is not matched by a corresponding change in species diversity. Usually the ratio D_{430}/D_{665} increases in deep water, as yellow pigments are better preserved in decaying plankton, but notable exceptions are found, as shown by the following averages:

Depth	Central Tyrrhenian, end of summer	Caribbean, offshore, June
0 m	4.48	4.6
10 m	5.10	4.2
20 m	4.56	4.7
50 m	3.42	6.1
100 m	3.60	3.7
250 m	2.96	

When settling down through layers rich in nutrients, populations of phytoplankton that were starving, especially because of lack of nitrogen, react immediately by an increase in green pigments and the value of the pigment index drops. But since the cells are now in very dim light, such rapidity of response is of no use.

The possibilities of the study of pigments in plankton research have not been fully explored. Unialgal cultures cannot exhibit changes in species composition through time but do exhibit changes in pigment composition as the culture grows older. Here, as perhaps in many natural situations, pigment composition is a more sensitive characteristic because it covers not only species composition but also the physiological state of their populations. Study of pigment changes can be readily introduced as practical work in elementary courses in ecology. I have referred to an experiment for illustrating succession. The following simple experiment is also worth some thought. During succession in mixed populations, either freshwater or marine, there is an increase in the pigment index.

Following the same process of succession, biomass appointment among the different species shifts almost continuously. If we culture three species separately, such as a green alga, a diatom, and a dinoflagellate, a similar trend in the change of the ratio D_{430}/D_{665} is observed in each culture. It increases in all of them, but the average level is different—the volvocale is greenest, the dinoflagellate is yellowest. If a system is formed by pooling all the species together, dominance of the different species changes during succession. In most experiments, the green organism develops faster at the beginning, followed by the diatom; in the end, the dinoflagellate may be more common, if it has survived at all. This sequence follows the order of increasing values of specific mean pigment ratio and is most significant. Any ecological succession or any process of self-organization of an ecosystem leads to an increase of this ratio. It is a very useful tool in ecological research, especially when we are looking for usable 'macroscopic' characters.

Pigment composition seems to be an equally useful character outside the pelagic environment. In aquatic plants with ribbon-like leaves (such as *Vallisneria*, *Posidonia*, and *Thalassia*), it is easy to recognize gradients in pigment composition which can be related to productivity in the different segments of the leaves; these gradients are easily measured by C^{14} fixation. In terrestrial plants similar differences exist: according to Bray (1960) the proportion of chlorophyll-b relative to chlorophyll-a increases going from the young to the older parts of a plant.

Particular attention has to be given to the development of devices to separate, extract, and evaluate pigments in continuously obtained phytoplankton samples. We have been experimenting with a simple device to extract pigment with acetone condensed on filters through which a measured volume of sea water was previously pumped. So far, results have not been very encouraging. But problems are mostly technical, and it should not be too difficult for the sophisticated technology of the day to produce a reliable device at a reasonable cost. Alternatively, the study of optical properties of natural waters could also lead to inferences about the quantity and quality of pigments in organisms suspended in them.

PRODUCTION AND BIOMASS

The analysis of plant pigments has often been intended as an indirect way to get estimates of production, biomass, or both. This is not a sound procedure and is even less so if we want to use pigment composition as an independent 'macroscopic' character to be compared with others. Production and biomass have to be estimated by special methods, however difficult. Some comments on the feasibility of such estimates are not out of order, however, because they may uncover some interesting problems.

Estimate of biomass from pigment concentration is hampered by the enormous variation in pigment content of the phytoplankton. Scientists have despaired of finding some conversion factor from chlorophyll to total dry weight. A small hope can be found in the fact that the amount of chlorophyll, or pigments in general, in plankton is related to the ratio D_{430}/D_{665}. Cells with a high pigment ratio—more yellow—have a lower pigment content than intensely green cells. When I was using Harvey units of pigments, which are proportional to optical density at 430 mμ, the convenient factor F for going from D_{430} to biomass (B) was found to be variable but roughly proportional to the pigment index squared. Thus, an estimate of biomass, not exact, but more reliable than if based only on units of plant pigments (Margalef 1960), was

$$B = C(D_{430})^3/(D_{665})^2.$$

The exact values of the constants in the expression, either C or the exponents, is immaterial because the interest lies in the suggestion of some general (allometric) relationship in the composition of pigments.

Production (P) has also been tentatively expressed as a function of quality of pigments (Margalef 1965b); a few data on freshwater lakes have yielded

$$P = C \, (D_{665})^{1.3}/(D_{430})^{0.6}.$$

If P is divided by B, the quotient, which expresses production per unit biomass, turns out to be proportional to some power (close to 3) of the inverse of the pigment ratio. This is only a rationalization of an empirically observed relation: the pigment ratio is negatively correlated with the production per unit biomass, or the energy flow (see Fig. 4). More evidence has been offered in another paper (Margalef 1963a). The ratio of production to biomass (P/B) is a fundamental 'macroscopic' character and has to be determined by independent methods. If pigment analysis can give an adequate estimate, so much the better, but that should not seduce us into complacency.

The introduction of the C^{14} method has prompted interest in the study of primary production, and many data are being accumulated. Nevertheless, their value is not uniform. There are still many problems involved. Short incubations give data subject to a strong daily rhythm which are difficult to integrate. Longer incubations cover an important recycling and the variance of replicates increases very greatly owing to divergent evolution of the bottle populations. Bacteria are also a problem, especially those which assimilate inorganic carbon in the dark; nor can the effect of the walls of the bottles be eliminated.

No less formidable are the problems involved in determining biomass. These are due in large part to the amount of detritic material present in every body of water which cannot be separated from living matter. It must be taken for granted that any evaluation of the very important ratio of production to biomass involves many indirect estimates and provisos and may be often quite inadequate. It has been said before that

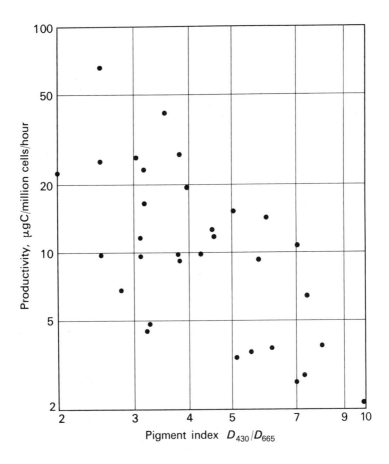

Fig. 4. Relationship between pigment index and productivity for a series of phytoplankton samples collected off Barcelona, 1965–66.

P/B is negatively correlated with the pigment ratio. There is also a negative correlation with species diversity (-0.72 in freshwater examples); see Figure 5 and Margalef (1964).

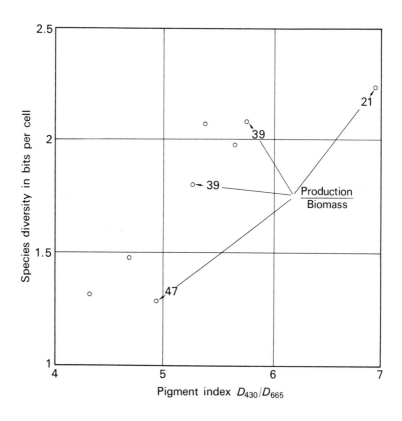

Fig. 5. Correlation between pigment diversity (abscissa) and species diversity (ordinate) in summer phytoplankton of artificial lakes in northeast Spain. When available, the ratio of production to biomass, in mgC/gC/hour, has been inscribed near the corresponding spot.

The Need for Technical Improvements

Various measures of diversity—diversity in species and pigments—in primary production and biomass represent important 'macroscopic' parameters of ecosystems. Recognition of this fact leaves the planktologist with a need for something that in the usual jargon of contemporary environmental science might be designated "a study of the feasibility, capacity, and development of an integrated plankton mission multistage operational facility, packaged in modular units of advanced technology." Indeed, we have many words, but it is necessary to devote to biological parameters an effort commensurate with that being dedicated to the measurement of physical and chemical parameters in the sea, or to anything in outer space.

The necessity of such an approach can be easily justified. It makes no sense to insist on the study of gradients in the sea if the only available measurements are those made in stations tens or hundreds of kilometers apart, and it is not possible to speak of rates of change on the basis of observations very distant in time. Temperature distributions, waves, and other phenomena have attracted some attention, but planktology has not yet tapped the possibilities of present technology. In order to continue research in a more intelligent way, an effort has to be made to perfect or develop systems for the analysis of the highest possible number of variables—systems as automatic as possible and with continuous recording. Considerable progress has been made in the field of measurement of physical variables, but planktologists are still worshippers of the venerable plankton net.

At present, months or even years pass before we are in possession of information on the results of any oceanographic expedition. A posteriori, one often regrets having passed over an interesting spot that, judging from conditions in neighboring stations, looked particularly promising for understanding some special phenomenon. It would be convenient to have an immediate and continuous representation of the properties of the pelagic ecosystem, with the possibility of changing the path of exploration, to maximize information in relation to particular problems. Echo sounders would be of no use to fishermen if the readable records were available to them only after their return to harbor; oceanographers too need information about the structures through which they are sailing.

It is possible to pump water from selected depths or along a sinusodal path and continuously analyze salinity, pH, oxygen, and selected nutrients. Sensors in the inlet pipe can record temperature and light. Optical counters (with photo-multipliers) can count big particles and, without the light in the counting chamber, luminescent organisms. Dimensional particle counters give a spectrum of the components of the seston. An automatic extractor and analyzer of plant pigments is feasible and we can even look forward to the construction of equipment allowing for the semiautomatic measurement of the capacity for inorganic carbon fixation. Echo sounders working at different frequencies may give information on distribution of larger animals. Simultaneous gathering of information multiplies its value, and data could be properly recorded, with reference to time (or position) and depth.

HANDLING INFORMATION AND THE IMPORTANCE OF DERIVATIVES

Very interesting problems arise in handling the wealth of information about which we have been dreaming. Some of the problems have been examined using as models data obtained in a discontinuous way or, if continuous, concerned with only a few parameters.

A possibly fruitful way to get an adequate knowledge of the ecological organization of the pelagic ecosystem would be to study continuously the correlations among the different parameters as the values are collected along a transect. One can start on the basis of previously observed correlations and, assuming continuity or constancy in the trends and conservation of the correlations, predict the configurations to be found ahead. The agreement or difference between expectation and observation may be a test of the understanding of planktonic organization and a means of improving such understanding. It is perhaps expedient to say this again in more detail. Suppose we had developed a series of regression expressions, estimated from several physical variables the concentration of chlorophyll, and found that the estimate fit well with the concentration found at some spot. Assuming a continuity in the observed trends, we then estimate physical parameters at the next spot ahead and go on to observe them, checking empirical data against the predictions. If the correspondence is good, we try to predict parameters of the populations. If everything fits well, we accept the consistency of our general schema.

It is more important to compare trends or gradients (that is, derivatives) than to compare point values. These gradients are necessary to understand the organization of pelagic ecosystems because the basic assumption in our general approach to the problem is that the composition and properties of plankton in neighboring regions are interdependent, the result of dynamic exchange phenomena. In the same way, but to a higher degree, interdependence exists between different regions of the body of an organism which are functionally and morphologically (that is, spatially) related. For example, there is no reason to foresee a correlation between phosphate concentration and phytoplankton density: the original phosphate level may have been low and thus limit the development of plants, or the low level may indicate depletion by plants. The rate of increase of phytoplankton may more reasonably be correlated with the rate of depletion of phosphate. Gradients in species diversity may be related to gradients in productivity much more satisfactorily than single values of diversity to single values of productivity: remember that progress in physics was possible only when derivatives were introduced.

Similarly, suppose that some way of estimating productivity based on regression is being tested. It does not seem appropriate to express production, which has the value of a derivative with respect to time, exclusively as a function of a number of parameters like temperature, phosphate concentration, light, and so on. Some parameter containing a reference to time and history should be logically included in a regression formula giving an estimate of production. Both diversity and pigment ratio meet this criterion; although dimensionless, these parameters are measures of complexity or history and carry with them some reference to previous time.

CORRELATIONS

One necessary step in ecological analysis is the study of correlations. Sometimes requiring a preliminary transformation of the data, correlations are computed between different parameters. Inspection of the matrix of correlation coefficients can suggest the computation of multiple regression equations, which allow an estimate of the value of some parameter as a function of others. Such relations are important as evidence of how parameters are linked to others, if they are linked at all.

Regression equations are just statistical expressions and it is not fair to ask dimensional consistency from them. As I have just suggested, however, it seems too much to accept the result of operations with instantaneous and point values as estimates of values that are derivatives.

Statistical correlations between parameters are not always very strong, but if the data are plotted on charts, congruency in patterns may show a relationship. This has been recognized by ecologists, who use maps extensively but statistical correlations rarely. Maps, of course, include gradients (that is, derivatives). The trouble with the ordinary procedure of

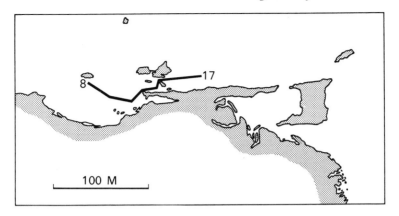

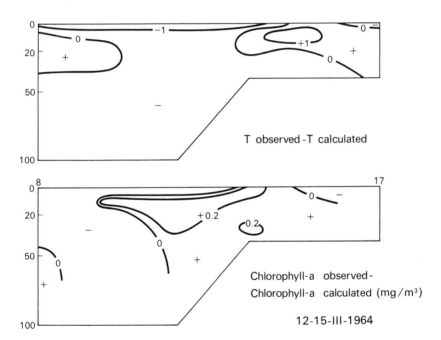

Fig. 6. Above, the coast of northeast Venezuela with the transect 8–17. Below, profiles from station 8 through station 17 with the differences between observed and computed values for temperature and chlorophyll-a concentration. Values were computed from multiple regression expressions.

statistical correlations is that the values are taken out of context of space and time to be mixed, so to speak, in a hat, a process which destroys all structure. With this procedure it is no wonder that scarcely any notion of organization emerges from statistical studies. If the values of a parameter are estimated as a function of others and (after actual values are observed) residuals are calculated and plotted, it may sometimes be found that residuals are not distributed at random, but show continuity. As an example, in Figure 6 residuals for temperature and chlorophyll-a are plotted along a section from station 8 to station 17 in the Caribbean (north of Venezuela). Temperature was estimated as a function of position, depth, and time; chlorophyll-a as a function of phosphate, nitrate, and light. The values plotted in the sections are the differences between the estimated values and the empirical values of the variables. Continuity of distributions is obvious. Also evident is a certain opposition between the two distributions: places with excess chlorophyll come close to places where temperatures are too low. Some causal connection might be sought, but in the original matrix of correlation coefficients the correlation between temperature and chlorophyll concentration is not significant (-0.06 for 577 pairs of data).

The configurations in space derived from the mapping of residuals reflect the distribution of stresses resulting from lack of equilibrium between environment and population. It can be assumed that such departures from ordinary relations are especially frequent under recently changed conditions or in the first stages of succession. Any extensive use of correlations, even with the inclusion of derivatives, must be considered a first step, giving a construct that must be incorporated into a larger frame. In other words, we may have rather strict correlations according to the geographical area, according to the season, and so on, but there is regularity in the way correlations change from place to place. Continuity of the residuals in space means organization; in time it is an expression of succession.

Taxonomic Evaluation of Samples

The propaganda against old and reliable taxonomy has been very effective; today, people able to identify organisms are distressingly scarce. Taxonomy is necessary even at a 'macroscopic' level, not only in computation of diversity but also in the study of similarity between different samples of ecosystems obtained by different methods.

The problem of characterizing types of ecosystems or communities by the use of groups of species is analogous to the problem of recognition of patterns by a machine (Margalef 1966). Machines that can recognize and learn to discriminate patterns and figures achieve this by using the spectra that express the degree of superposition of the image with every one of a set of uncorrelated templates. The interesting suggestion is that an adequate discrimination can be achieved midway—that is, by using neither all species present nor too small a number of groups, but instead choosing an appropriate number (for instance, around 15) large enough to achieve good discrimination and reveal spatial heterogeneity and small enough to make the operation simple. The groups can be defined by using the methods developed by numerical taxonomists; since it is possible to recognize clusters of species that are found most frequently together, the presence of different species of a group is therefore redundant. I have worked along these lines on Mediterranean phytoplankton. It seems that the pattern of association changes from place to place and from time to time. Just as we have sets of correlations peculiar to areas and times, to be included in a wider system, we find here that the association between species has only a limited value and is part of wider regularities.

Any automatic program being fed cards or punched tape

with recorded data about species identified in the course of an investigation could systematize the pattern of clustering between species, and on this basis evaluate spatial differences. At the same time, it could reevaluate the data continuously to see if the same clusters are still applicable. If new rearrangements of species are advisable it could propose them. In other words, departures from an old regularity should always be detected, but they should be recognized in the context of regularities of a higher order.

SUCCESSION AND HETEROGENEITY

I have examined elsewhere in some detail the problem of succession in plankton and the spatial differentiation of populations related to succession (Margalef 1958, 1962*b*, 1963*a*).

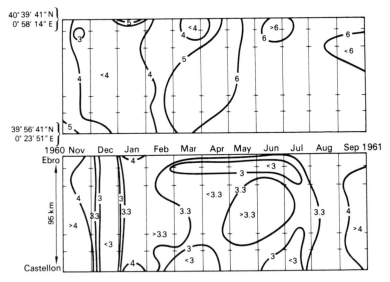

Fig. 7. Values of the pigment ratio D_{430}/D_{665} (above) and of species diversity measured in net samples (below) for the phytoplankton of surface water along a coastal stretch in the western Mediterranean from November 1960 to September 1961. Positions of the endpoints of the transect are given at left. Diversity is lower in the periods of maximum productivity.

The succession in plankton is a paradigm of the model of succession given in the second chapter. It starts with mixing and fertilization and proceeds as far as stabilization of water allows for a vertical differentiation and a certain amount of stable organization. Primary production per unit biomass decreases during succession and diversity at every level increases, if one takes into account the existence of a spectrum of diversity. In advanced stages of succession diversity may be quite low in small volumes of water. In Figure 7 an example of changes in diversity in Mediterranean water is presented; Figure 8 provides an example of change in the pigment ratio D_{430}/D_{665} in experimental successions in the laboratory which duplicate very closely natural plankton blooms. This example is intended as a suggestion of the vast potentialities of an experimental approach, either in the form of simple cultures or in the form of chemostats or flow cultures, in which exploitation

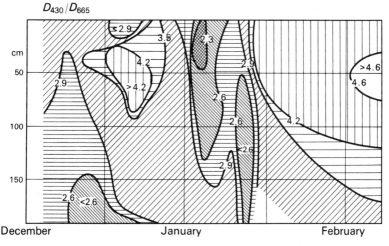

Fig. 8. Variation of pigment ratio D_{430}/D_{665} in a culture vessel 200 m high subjected to a light- and turbulence-gradient in simulation of natural conditions at sea. In the period under study two outbreaks were induced by fertilization and mixing; the first one with dominance of *Skeletonema* and the second with dominance of a small green alga.

can be changed by regulating flow. The only serious drawback to the experimental approach is the presence of solid walls.

Real successions at the level of phytoplankton are co-extensive with simple waves of population at the level of many zooplankters and are synchronized with internal rhythms of fishes and other animals. Breeding cycles in populations of animals are established in such a way that the maximal number of young and the highest exploitation of phytoplankton is found in the early stages of phytoplankton succession. This results in a sort of vicious circle: because of being subjected to strong exploitation, early stages of phytoplankton succession last longer, or the population even becomes rejuvenated. There is a constant interaction between plants and animals in regulating the speed of phytoplankton succession. On the other hand, quality of food, which is different in each stage of succession, has a tremendous impact on zooplankton development and evolution. As succession proceeds, food, at first divided in very small particles, tends to evolve toward a different distribution. The total amount becomes smaller and is distributed in a much reduced number of larger particles, many of which are motile and, sometimes, toxic. Therefore, it pays for animals to switch from passive feeding to a macrophagous habit. The high proportion of macrophages and the elongation of food chains are general characteristics of very mature oceanic ecosystems, the only ones in which giant squids and sperm whales are at home.

It can be safely affirmed that no two small volumes of the pelagic ecosystem are congruent. All the water of the ocean forms an ecosystem with no replicate parts; there are big structures containing smaller structures and so on, down to a single drop of water. There is nothing strange in this. Much simpler systems, for example, physical systems such as the atmosphere and the hydrosphere, present an analogous complexity of structure. A hopeful perspective derives from the fact that the distribution of macroscopic properties, such as diversity or primary production per unit biomass, helps to represent and summarize such a tremendous range of diversification and organization. In Figures 9 and 10 a few examples of medium size structures are shown. Probably the hypothesis holds everywhere that the less mature ecosystem feeds the more mature structures around it.

An uneven demographic structure of population can be recognized for animals that disperse and move through ecosystemic structures of heterogeneous maturity. Young animals are relatively more common in less mature systems where food is in surplus, and the apparent death rate is higher in such places simply because there is a continuous flow of animals, as they grow older, to more mature areas.

Populations of the sardine (*Sardinia pilchardus*) along the Mediterranean coast of Spain afford a nice example. Planktonic ecosystems, going south from the mouth of the Ebro

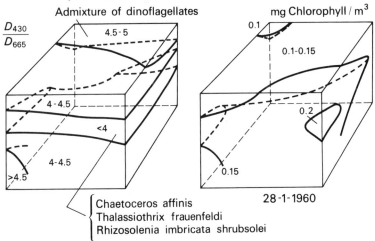

Fig. 9. Distribution of some properties of plankton in a block one square mile in area and 60 m deep in the western Mediterranean. At left, pigment ratio; in the space where pigment ratio was below 4, population was dominated by diatoms; in places with pigment ratio above 4.5 at the surface there was a strong admixture of dinoflagellates. At right, distribution of chlorophyll-a.

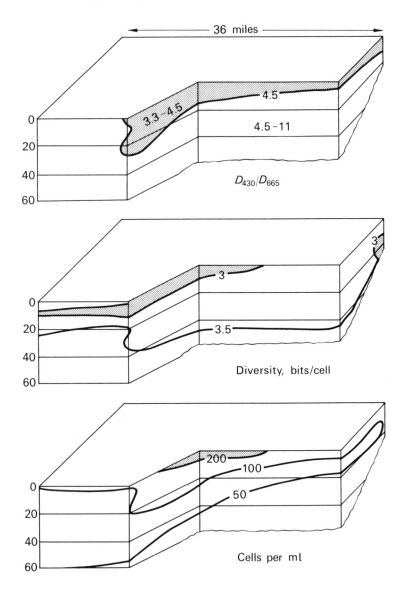

Fig. 10. A diatom bloom (*Nitzschia* dominant), in the western Mediterranean, May 1962. The spaces of maximum cell concentration are associated with a low species diversity (measured in sedimented samples) and a low pigment ratio.

River, show an increase of maturity in all essential 'macro-scopic' characters. In the same area, the distribution of age classes in the populations of sardine show corresponding local differences, with a predominance of younger-year classes in the areas where the whole pelagic system is less mature and with an older demographic structure in the areas of more mature characteristics. Another, more trivial, example of this very general regularity is the vertical distribution of demographic structure in all pelagic animals. Populations are younger near the surface, where the pelagic ecosystem pumps more energy through the biomass.

4

Evolution in the Frame of Ecosystem Organization

SUCCESSION AND EVOLUTION

This subject would be best described by the title, but only the title, of a delicious book by G. E. Hutchinson, *The ecological theatre and the evolutionary play.* I presented the basic points of view sustained here in a short communication to the Fifteenth International Congress of Zoology, held in London in 1958 (Margalef 1959). This paper passed unnoticed, so it is not superfluous to present the same argument again, including some further development.

Books on ecology sometimes include a chapter on ecology and evolution which deals principally with isolating mechanisms, the action of environmental factors on the frequency of mutations, genetics of populations, and some comments on the wonderful adaptations revealed by the field of ecology. Here I want to adopt a different point of view and a different reasoning, perhaps with certain similarities to the paleontologist's way of thinking. In both areas, that is, in paleontology and ecology, the picture is sketchy and oversimplified and there may be actual danger of being led to make sweeping, though incorrect, generalizations. Many of the supposed mechanisms and pathways of evolution must be considered as tentative, for geneticists have neither confirmed nor dismissed them.

Natural selection is not as wasteful as it is assumed to be in popular books because it uses energy that would be lost anyway in keeping an ecosystem running. The energy gates at the places where species interact—or where they interact with environment—are the organs by which selection is achieved and evolution occurs, the rate of evolution depending on the efficiency of the gate. Succession in any self-organizing system involves the substitution of some piece of the system by some other piece that allows the preservation of the same amount of information at a lower cost, or the preservation of more information for the same cost. In this context, information is anything that can influence and shape the future, and the cost is represented by the energy used, which amounts, practically, to the energy entering the ecosystem as primary production.

Evolution cannot be understood except in the frame of ecosystems. By the natural process of succession, which is inherent in every ecosystem, the evolution of species is pushed—or sucked—in the direction taken by succession, in what has been called increasing maturity. The implication is that in general the process of evolution should conform to the same trend manifest in succession. Succession is in progress everywhere and evolution follows, encased in succession's frame. As a consequence, we expect to find a parallel trend in several phylogenetic lines which can also be recognized as a trend realized in succession. For instance, a decrease in the ratio of production to biomass and an increase of the efficiency and specialization should be common trends in phylogenetic lines. Such iterative paths in evolution have given rise to words and so-called theories like orthogenesis, a non-explanatory term meaning only 'regularity in direction'. Orthogenesis results from the combination of what is possible in the already acquired organization of the genetic line with what is required by an ecological theater that is always shifting in the same direction.

Actually things can never be so simple. Evolution cannot be wholly encompassed in the slow pace of succession. Without doubt, many lineages have passed along their phylogenetic history from one type of ecosystem to another. Other modes of evolution may represent passage from one trophic level to another, as probably has happened more than once in the evolution of pelagic crustaceans. This may be a rather discontinuous process. Nevertheless, I think that what is more important, and what has not been sufficiently stressed, is that the elementary steps of evolutionary change are always constrained by the organization of the ecosystem, always caught in a similar process of self-organization. Perhaps some exceptions exist. In an ecosystem where trophic layers are being added, the lower trophic levels are rejuvenated, kept by exploitation in a condition of low maturity, and species in those lower levels are not subject to the usual momentum of succession. They may become stagnant. Perhaps this is the reason why so many primitive types (silicoflagellates, radiolarians, etc.) are still found in the plankton relatively unchanged, or as zooxanthellae enslaved by animals.

There is another facet to the subject of prospective evolution. New empty spaces are continually produced—places where the ecosystems are altered or destroyed—where succession starts anew. The mosaic of the biosphere shows all possible states of succession. Evolution going in the other direction, producing opportunistic or pioneer species in ecosystems of low maturity, is less documented, probably because it is more rapid and its results are more difficult to understand. Tempo and mode in evolution appear linked to basic properties of ecosystems, in a most fundamental feedback circuit. The properties of an ecosystem depend on the organisms of which it is formed, but the evolution of such organisms is under the control of a process of self-organization operating in the whole ecosystem.

CONDITIONS OF SELECTION IN DIFFERENT STAGES
OF SUCCESSION

Characteristics of ecosystems in different stages of maturity explain differences in evolutionary opportunity. Among the characteristics that should be considered are the following:

Fluctuations

These are intense in young ecosystems and less important in mature ones. Population fluctuations are more detrimental to parasites than to their hosts and, in general, to species very dependent on others. They also shorten the food chains. Strong fluctuations select for prolific species; relative constancy in numbers leads to restriction of numbers of offspring, with protected embryos, parental care, territorial behavior, and so on. (Compare clupeids with coral fishes, euphyllopods with crayfishes, and birds nesting in open sites with hole-nesters.)

Species that thrive in fluctuating systems must have a high rate of potential increase, and in consequence the flow of energy through their populations is high. For such species the sum over long periods of time of the squared differences between multiplication and mortality, $(b-m)^2$, may be a measure of their creativeness; in fact, such species are relatively expensive from a thermodynamic point of view. As a consequence, their evolution may be more rapid than that of species evolving in mature or maturing ecosystems. The cost of natural selection, in the sense of Haldane (1957) is strictly linked to the efficiency of the energy gates in the ecosystem; the cost is high or low depending upon whether energy flux per unit biomass is high or low.

The importance of fluctuations in population genetics is generally accepted. One should perhaps keep in mind that organisms ordinarily used in genetic research (such as *Neurospora*, *Tigriopus*, and *Drosophila*) are rather similar to pests or cultivated species—that is, to species characteristic of systems of low maturity. Information on the genetics and evolution of species well-adjusted to mature systems is almost completely lacking, especially because species with a long life span and low fertility are not preadapted to the geneticist's laboratory. Nevertheless, different species of *Drosophila* show adaptation to ecosystems of different maturity and comparative study of them can provide a glimpse of the relative importance of the mechanisms at work in different kinds of ecosystems.

Dispersal

Species that disperse easily can maintain a strong genetic flow between distant populations. In general, facility for dispersal is related to the number of propagules produced and diminishes as maturity of ecosystems increases. As I have noted previously, many plants from grasslands are anemochores; forest dwellers either have no special means of dispersal or are dependent on animals bound to a limited area. From this point of view the comparison between planktonic and benthic species is dramatic, and comparative speciation in pelagic and demersal fishes is probably related to their respective facilities for dispersal.

Predictability of environment

Direct reaction of organisms to environmental change is most useful if the environment is being altered in an unpredictable way. In more regular and organized environments, rhythms may be better governed by indirect responses— responses to especially regular factors lacking ecological importance in themselves but associated with other factors of great importance in life—or else by endogenous rhythms. Endogenous rhythms have the power of anticipation, and they nullify time lag, thereby increasing general stability. A similar shift in components can be detected in the more complex forms of behavior: in the relationship between inherited or stereotyped behavior and in environment-oriented or developed components of behavior. Stereotyped behavior, with an important inherited component, is cheaper and eliminates any period of adjustment, but requires a predictably organized environment. One is led to suppose that the fraction of such stereotyped behavior must be greater in the very mature ecosystems.

Turnover

In related species, turnover is higher in forms located in less mature systems. One example is afforded by the brittle stars of the genus *Amphiuma*, studied by Buchanan (1964). In the less mature "*Echinocardium-filiformis*" community, located on silty sand at depths of 15 to 70 meters, the proper species is *A. filiformis*, which lives for 2 to 4 years, feeds on poorly selected bottom detritus, and has an oxygen consumption of 0.06 milliliter per gram of wet weight. On fine sandy silt at depths of 70 to 100 meters, the obviously more mature "*Brissopsis-chiajei*" community is found; in this community, *Amphiuma chiajei* is found, which lives for more than 10 years, feeds in a somewhat specialized way, and shows an oxygen consumption of 0.01 milliliter per gram of wet weight, measured under the same conditions as for *A. filiformis*.

Species in stages of low maturity or those belonging to the lower trophic level have in general a shorter life, and there is sometimes a possibility for periods of latent life. If life-span is shorter than the principal periods in environmental cycles, successive generations live under different conditions and the environmental cycles act as a successive series of filters of different selectivity. The result is that species may attain a plasticity (cyclomorphosis or at least physiological plasticity) with the power to adjust to a large range of conditions. Such plasticity is obvious in freshwater copepods with a 2 to 3 month life span (*Tropocyclops, Eucyclops*) and seems to be associated with a rather poor geographic differentiation of genotype. In marine plankton there are many monotypic genera (Friedrich 1955) which often possess notable pheno-genetic plasticity.

If life-span is adjusted to cover one or more environmental cycles, successive generations develop under equivalent conditions, and a stereotyped, canalized development may be selected. Conditions may change during the life of each individual, but all individuals experience comparable patterns of change. The freshwater copepods of the *Cyclops strenuus* group and a large number of the diaptomids are appropriate examples. Life-spans of approximately one year and, in place of intraspecies plasticity, a strong geographical variability of undoubted genetic basis are observed.

Feeding and competition

Mature stages contain a higher proportion of stenophagous animals and there is presumably more interdependence of species based on the exchange of ectocrine substances with special biological activity. There is a biochemical evolution of metabolic deficiencies (Florkin 1949), and such evolution runs parallel to the increasing availability of the most varied compounds; that is, regular connections with succession can be expected. It would be interesting to draw, for different stages of succession, a detailed report of the diversity of organic compounds available at every stage. Such diversity would probably increase with maturity, but there is not yet enough information available to be certain.

In plankton, indiscriminate filter feeders are at an advantage in the initial stages of succession, but, given the usual properties of phytoplankton and other food organisms, it can be shown that in later stages of succession there is an advantage in evolving from microphagous behavior to a more selective hunting behavior and in concentrating increasingly on larger food items (Margalef 1967*a*).

Initial stages, characterized by strong competition for food and a superabundant fecundity, often show the exclusion of congeneric species. Biochemical integration in mature ecosystems mitigates competition for dominance and frequently leads to relations mediated by ectocrine substances; this is only possible in water when a certain degree of stability has been achieved. If such ectocrine substances are characteristic of genera rather than species, congeneric clusters of species may be favored (*Goniaulax*, *Caulerpa*). It is probable that such alleviation of intraspecific competition may favor a further splitting of species. A serious study of the ecology and evolution of freshwater desmids or marine dinoflagellates, always more diversified in very mature systems with a low energy flow, would be rewarding from this point of view. The dynamics of the populations are slow and a hundred congeneric species are sometimes found together.

In some cases it may be useful to study the problem of interspecific competition in relation to the species' adaptations to ecosystems of different degrees of maturity. This could be done, for example, in the laboratory flour-beetle so extensively studied by Thomas Park. So far, these studies have typically been conducted in a laboratory universe which is rejuvenated at monthly intervals by renewing the flour. If the flour is not renewed, however, information accumulates in the form of partially consumed carcasses, exuviae, fecal pellets, and glandular secretions such as ethylquinone. The phenomenon of competition in this more mature system has not been studied.

In very mature ecosystems, a multiplicity of links and a general slowing down of dynamics result in less brutal and more subtle competition. The success of species is linked to multiple and delicate relations. Species adjusted to a way of life that depends on the persistence of the whole ecosystem may evolve to practice some sort of ecological altruism (for example, trees giving bread and board to insects, in the form of galls). Such characters could be considered useless and even detrimental if the species are considered separately. The apparent paradox of species with a lower rate of increase outcompeting species with a higher potential rate of increase is central to the problem of evolution of competition.

CHARACTERISTICS OF SPECIES ADAPTED TO DIFFERENT STAGES OF SUCCESSION

Summarizing what has been said, the comparative study of different stages of succession may uncover many mechanisms that explain why they are the source of species with peculiar evolutionary characteristics. The extreme types can be characterized as follows:

Initial stages

These are characterized by species of short life-span which leave numerous descendents and have long-range ways of dispersal. These species are capable of developing adaptive cyclomorphosis and local adaptive modifications and maintaining a certain genetic unity. They are euryoecic, adapted to changing conditions in time and space (Levins 1962–63). They compete for dominance; they are also capable of rapid evolution. The most important criterion, opportunity, encourages a dynamic type of selection (cladogenesis, aromorphosis). The multiplicity of names such species have received—fugitive, pioneer, opportunistic, prodigal, apocratic, or kinetophilous (Erdman 1963)—testifies that their nature has been recognized and that they have awakened interest.

Mature stages

These contain species with a longer life-span which are isolated in small breeding communities, produce a small number of descendants, and have poor possibilities for dispersal outside the immediate neighborhood. Biochemically and behaviorally they are very well integrated into the system; their characteristics include a territorial instinct and a need of active substances (e.g., vitamins) produced by other organisms. Canalized development and strong stability, both morphologic and genetic (ready elimination of 'creative' variants), are also characteristic of these species. Competition may shift to the level of the genus. Efficiency, the key criterion, leads to a stabilizing type of selection (anagenesis, telomorphosis, orthogenesis).

All intermediate types may exist. The main point is that slow and well-documented evolution shows a general trend from species of the first type to species more adjusted to an increased maturity.

LIFE TABLES

Differences between species are best described by survivorship or mortality curves. It is a fair generalization that species of less mature ecosystems or of lower trophic levels have a diagonal (sharply falling concave) survival curve but that species of more mature ecosystems or of higher trophic levels tend to have a rectangular curve. The curves differ not only in shape but also in total dimension, since species of more mature ecosystems may have a longer life. Distribution of the reproductive period over the life-span may also be typically different (Fig. 11).

What has been said before can be expressed in another way as a trend in the evolution of the shape of survivorship curves; the slope decreases over most of the range and the sharp drop at the end of the rectangular curve appears and may be shifted.

Decreasing the slope of the curve means a reduction in mortality. Protection of young, larger size, low metabolism, development of defensive mechanisms, and passage from an indiscriminate type of feeding to a more selective one all

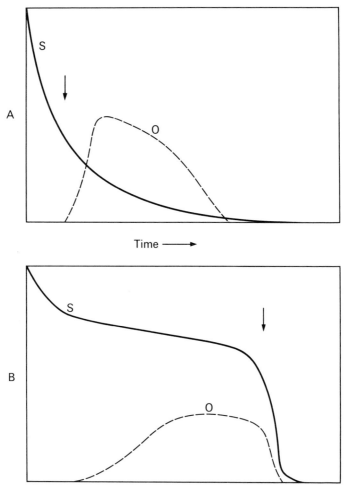

Fig. 11. Survivorship curves (*S*) and age-specific fecundity curves (*O*) for two idealized species: *A*, a species of an ecosystem of low maturity or of an inferior trophic level and *B*, a species of a system more mature. The arrows indicate the ages at which natural selection is stronger.

contribute to the decrease in mortality rate and longer life. The fight to eat and not to be eaten, coupled with increase in size, often involves change in trophic level and is a part of this trend.

It is not by chance that all the most advanced examples of defense—mimicry, animal poisons, elaborate symbiotic relations, and complicated territorial behavior—are more frequent in the most mature ecosystems. They characterize the coral reef in the sea and the tropical rain forest on land. The cause is not the "life-promoting force of the tropical sun," but a long evolutionary opportunity in a very mature ecosystem. In other very mature systems in rather cold environments, evolution has been going on in the same direction: abyssal benthic fauna is composed of large-sized, sparsely diffused species with low metabolic rates, slow reproduction, elaborate ways of protecting offspring, and so on, as is typical of species in the tropical rain forest. The same can be said of true cave communities, including, for instance, beetles with long lives, low metabolism, and small numbers of descendants born at a very advanced stage of development.

The rectangular survivorship curve indicates a rather strong physiological limitation of life-span. The trend toward immortality must have limits, and should not be pursued immoderately. Such a rectangular curve conforms to the theoretically most efficient link in a food chain, in which prey is eaten just when it is about to die naturally (Slobodkin 1961).

The high mortality represented by a diagonal survivorship makes more deaths available for the workings of selection; thus, evolution can proceed faster in systems of low maturity.

The points of maximum mortality in the survivorship curve are the ages at which the weight of natural selection is most important. In the rectangular curve natural selection works at the end of the determined life-span, perhaps lengthening life or delaying the reproductive period; this causes elaboration of characteristics associated with later stages of life (telomorphosis). On the other hand, in species where the highest mortality is concentrated in younger stages, selection must operate in these stages. Remember that in populations of species spread over systems of heterogeneous maturity, younger stages are associated with the less mature segments, where they are subject to many factors of mortality.

Selection in species with diagonal survivorship curves most probably favors precocity in reproduction, which compensates for the small number of individuals that are preserved in later stages of development. Perhaps such older stages are dispensable; this would shorten life. At least, this must happen often under some changes in environment. This condition is called 'neoteny'. Associated with neoteny are parthenogenesis and other forms of nonsexual multiplication, which are frequent, for instance, in parasites.

The planktonic appendicularia and salps are probably neotenic forms of sessile tunicata, as cladocera may be neotenics of conchostraca. In algae, the morphologic series from flagellate cells, which pass through gleocapsal, cellular, and filamentous types (linked with the passage from planktonic forms to benthic ones and repeated in different groups), exemplifies a regular trend in evolution. Under certain conditions, regression from a filamentous or a sessile stage to free-swimming forms, as in *Prasinocladus*, may be considered neoteny.

Changes that impose a regression of ecosystems to a less mature stage, or the opening of new spaces to colonization, create new opportunities for the development of new species; such evolution does not take a slow and regular path but proceeds through neoteny or other nonhabitual or poorly understood evolutionary paths. In the words of Beer (1961), the energy made available by the disturbance is used to make evolutionary advances.

EVOLUTION AND SUCCESSION IN PERSPECTIVE

In the course of geologic time there has been constant rearrangement of ecosystem organization; very old and mature ecosystems are destroyed and new spaces are opened to colonization. Some ages in the history of the earth have witnessed more drastic changes than others, and interpretation of such discontinuity, from the biological point of view, in its exaggerated form has led to the theory of periodic cataclysms of Cuvier and others.

The important fact is that under conditions of relative stability in the environment, the trends in ecosystem succession and species evolution have been congruent and conform to the outlines we have sketched. Evidence from the development of mesozoic forests, from the evolution of reptiles, and from the development of marine faunas substantiate the basic hypothesis that trends in evolution have followed a blueprint somehow implicit in a process of self-organization of systems. Increase in size could be explained as the result of a selection pressure minimizing the danger of being eaten and the increased probability that big individuals have of leaving offspring. These inherent advantages accrue as a bonus to bigger variants (Kurten 1953). But only by considering evolution in the context of the ecosystem can we understand the selective value of any change that leads to a decrease in the energy flow necessary for the maintenance of a unit of information. This reduction occurs not only in the species but also in the whole system of which the evolving population is an expendable part. In other words, selection is controlled by the cybernetic mechanism at the level of ecosystem.

It is very important to compare evolution and succession now and in the past. Our knowledge of these two processes is unbalanced in opposite ways. In the present we can study the mechanism of succession but we have barely glimpsed the long-range, constant pressures working in evolution. For the past we have well-documented information about phylo-genetic series of foraminifera, cephalopoda, brachiopoda, vertebrata, etc., which assures us of the reality of definite and common trends in different lineages; our knowledge of suc-cession in the past, however, is at best sketchy. If we are allowed to combine, in an intellectual synthesis, evolution yesterday and succession today, the emergent image would be of considerable solidity. The working principles may be applicable to any cybernetically regulated system which is composed of reproducible and expendable parts able to evolve; thus, these principles may apply to any equivalent of species and ecosystems anywhere in the universe.

The starting points for a phylogenetic series in which evolution is supposed to have been rapid and conditions of preservation of remnants rather unfavorable seem to coincide with immature ecosystems. In such a stage evolution could be rapid, in accordance with the high price paid for it, and it is clear that transient environments are not good places for the preservation of remnants of the past. It can be said only half facetiously that the effectiveness of mature ecosystems in transferring information to the future includes the ability to transfer even dead information, in the form of structures and fossils.

Historical events may have led to a less than perfect equi-librium between succession and evolution. The strategy of life, in breaking new ground, outstripping adaptation of other lineages, and interfering with other phenomena, may have introduced tensions with secondary importance to evolution. Such imbalance may be exemplified by two cases. One is colonization of aquatic environments by plants evolved on land. Such colonization has been less important in the sea than in fresh water. In any case, the success of plants in the new environment may be related to the fact that old aquatic animals which were potentially herbivorous were not adapted to utilize the structural glucids of flowering plants or were at least much less able to do so than terrestrial animals. Marine sea grasses, as such, are rarely used as food by marine animals, and in the cycle of the respective ecosystems most of the plants fall as detritus before being utilized. It is interesting to note in passing that the most successful exploiters of the sea-grass meadows are animals of terrestrial origin, like the manatee and turtles, which arrived, so to speak, after the plants. A second example, already discussed before, concerns the relatively low ratio of animal to plant biomass in terrestrial ecosystems.

Climate and Evolution

To add a geographical dimension to the picture, it can be said that in a rather constant climate ecosystems can attain stages of high maturity whereas areas subjected to fluctuating climates may have experienced destruction of well-organized ecosystems. In the latter, new lineages may arise by neoteny or other more or less heterodox evolutionary mechanisms. Climatic fluctuation in this context refers to longer periods than the yearly cycle. Glaciated areas are, for instance, an example of fluctuating climate.

Speculation about such possible relations between climates and evolution is not new. Mathew (1915), in a long paper titled "Climate and evolution," and Taylor (1934), writing with more concrete reference to human evolution, assumed correctly, I believe, that important evolutionary changes occur in areas of climatic fluctuation and then spread peripherally toward more stable climates where the lines are caught in the general trend of slow evolution. The tropical belt may not necessarily be a source of new lines but rather a refuge of types that have originated elsewhere and then migrated. On this point there is no agreement because certain distinguished biogeographers believe that the tropical zone has been the place of origin of many groups. It must be recognized that in mature environments some evolution may take place against the general trend (in the direction of increasing energy flow), either in the colonization of small subsystems of low organization (soils, interstitial water) or by rejuvenation under pressure of exploitation exerted by other trophic levels.

But it is difficult to understand how species adapted to systems of low maturity can arise in large numbers in a place from which old mature ecosystems have not been wiped out and where there is no stimulus in the form of environmental change forcing energy through the system and making it available for evolution. I must confess again that I am much in sympathy with Mathew's hypothesis.

Man and Evolution

The appearance and evolution of man and his cultures were probably assisted by fluctuating climates, maintenance of systems at a low maturity by fire and other means, and by the poor exploitation of the plant resources implicit in the exceedingly low ratio of animal biomass to plant biomass in terrestrial ecosystems. Probably some neoteny associated with the reduction of maturity was involved since man remains, in general, a being adapted to rather low maturity systems, although differences exist among the prehuman and human groups.

The evolutionary play was going on in the evolutionary theater when as a part of the plot men entered, romping and stamping on the stage and bringing it almost to the point of collapse. The feedback between ecosystems and evolving species is obvious in the relations between man and nature. The second part of the circuit involves the impact of man on nature. In a previous chapter the consequences of human action have been described as a reduction in the maturity of ecosystems. Man is pumping energy through the ecosystem, thus rejuvenating the entire biosphere and opening new opportunities for evolution. The evolution of man has not been in the direction of passive adjustment to more mature ecosystems but is actively sustained through a regression of the rest of the biosphere. In this sense, technology and modern life are very wasteful. The amount of energy used to preserve a unit of human biomass seems to be increasing, but perhaps the amount of energy used to preserve a unit of information does not increase so much. Man is a historical accident, the creator of a stress that cannot be ironed out in the proposed model of succession and evolution. In this sense he is comparable to the development of trees with almost indigestible trunks.

THE TRIPLE CHANNEL OF TRANSMISSION OF INFORMATION

Now I want to go back to the first chapter and the subject of nature as a channel of information. Only a universe of energy could have no past. If there is matter, structures grow and differentiate and a past can be recognized and partially reconstructed. It is the problem of durationless nonmatter versus enduring matter (Fokker 1966). At one end of the spectrum is biblical chaos, a past without a past, because no matter exists to convey information. At the other end there is only information and no decisions—static information forever. We and the entire universe are caught in between, with the wonderful opportunity to enjoy creation. Some sophisticated people would rather speak of existential anxiety, but biologists in general and field naturalists in particular are really childlike and enjoy nature like a child playing in a mud puddle after the rain has given way to sunny skies.

A closer examination of the ecosystem, considered as a channel of information, allows one to separate three different layers, or subchannels. One is a genetic channel in replicable individual structures. Another is a truly ecological channel based on the interaction between different cohabiting species and expressed in the relative constancy or in the regular changes of their numbers. This channel is the one continually referred to throughout this book. A third channel may be called 'ethological' (because ethology is the science of animal behavior) or 'cultural'; it transmits what has been learned by individual activity or experience and is transmitted to future generations outside the genetic channel. This last channel had a negligible importance at the beginning of life but it is now increasing explosively. In it can be placed: formation of trails and burrows that are used by other individuals, accumulation of dead material, imprinting, imitative collective behavioral memory and formation of local traditions, and the legacy of tools and all cultural manifestations in man. Indulging in a figure of speech, one can say that it runs from the dead wood in the forest that preserves a structure to the wood transformed into paper in our libraries that preserves culture. And as books usually contain more information than tree trunks (or at least more readable information), it is clear that the conversion of trees into books will go on for some time.

An interesting subject of speculation is to estimate how the total information transmitted to the present time in the biosphere has been distributed in the three channels. Otherwise the separation of the three channels is devoid of interest.

The feedback circuits formed by interaction between species are a very expensive memory with a relatively limited capacity for storage. At the ecological level, contemporary forests do not convey much more information than mesozoic forests. Climatic fluctuations and other changes in the conditions of life have forced succession to start again and again in different places, and old information has not been saved in the form of feedback circuits in ecosystems; rather, species have been replaced by others and the ecological memory seems always to have played the role of an auxiliary memory of rather limited capacity. The opposite is true of genetic information. Although sometimes latent for long periods of time, it has preserved increasing amounts of structure able to influence the future. The genetic channel has undoubtedly enlarged more rapidly than the ecological channel. Surely the cultural or ethological channel has become tremendously enlarged with the development of higher vertebrates, and its increase reaches the proportions of an explosion with the

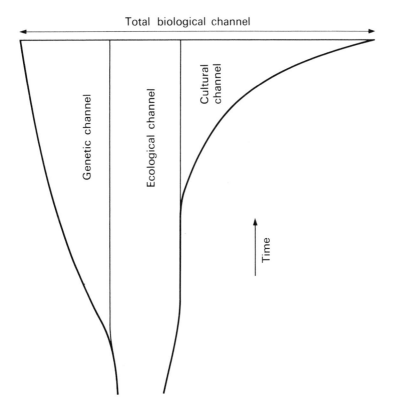

Fig. 12. An idealized view of distribution along time of information forwarded by life, in three channels.

advent of man. Now, if we project the relative size of the channels back in time (Fig. 12), the total channel is like a fan divided into three unequal parts: an ecological channel enlarging negligibly, a genetic channel enlarging considerably, and a cultural channel appearing later but enlarging explosively.

The Beginnings of Life

If in the course of evolution the genetic and the cultural channels have increased in size more than the ecological one, there is reason to ask if perhaps, in the origin of life, the ecological channel was relatively most important. In other words, we may speculate about whether the mechanisms proper to the ecological channel—that is, feedback between separate entities increasing at different rates as a way to maintain organization—are not, in fact, the ones closest to life and its beginnings.

From all the speculations of different authors about the origin of living systems, one is led to conclude that such primitive systems were more comparable to ecosystems than to present-day organisms. By this I mean that primitive systems, before they possessed a mechanism of molecular copying based on nucleic acid chains, could preserve some organization through feedback between different parts of a chemodynamic machine. This involved a sort of circuitry not very different from the one existing today between the different species of an ecosystem. If so, every primitive living thing was at once unique and original, with almost no replicate parts, just like an ecosystem—with apologies to the systematic phytosociologists! That is, these systems had a structure perfectly adjusted to local properties of environment and everywhere adjusted to the locally available materials more than to the available energy. This is a matter of choice and opinion; the authors most popular today believe that solar energy was not necessary and that primitive life had the form of material systems competing for energy. I am more inclined to Needham's (1959) view that living systems have always been energetic systems competing for materials. The distinction is important because transmission of information is more closely linked to the possibility of organizing huge amounts of matter than to the possibility of letting through high amounts of energy.

The beginning of species and of evolution can be placed at the first moment when mechanisms of molecular duplication were effectively used. Huge amounts of information could be copied—copying is very important, for it is cheap, like printing—and distributed. Instead of slowly assembling again and again a different organization in every location, there was the possibility of putting together rapidly, at a low expense, equivalent systems made of prefabricated pieces—that is, individuals. A small assortment of such pieces (species) was necessary, some to serve as primary producers and others to close the cycle of matter, regulating the respective numbers by feedback circuits. Thus, the genetic channel was born and information started to flow from the ecologic to the genetic channel, where it was better preserved and transmitted.

The hypothetical organism-ecosystems of the beginnings of life were assembled in place, in perfect adjustment to local conditions and subject to slow change: succession and evolution were one. Later ecosystems were organized on the basis of prefabricated pieces, and the relative numbers of each component in the total system were regulated, as has been described. But with such a model of construction, adjustment to local conditions can never be total: with pieces cut to standard sizes it is not possible to cover exactly any arbitrary surface. The result is that all species in the frame of an ecosystem are subjected to some stress, at least to the stress resulting from the succession or self-organization of the ecosystem. In other words, the primitive organism-ecosystems were probably more flexible than contemporary ones and species could not be separated from the primeval 'soup' or 'cake'. The introduction of the relatively rigid mechanism of genetic transmission, a successful tool of life, led to the separation of species inside the broth.

A proper subject of ecology is the study of the increase and preservation of organization at the ecosystem level; the fashion of the times leads one to emphasize some aspects of cybernetics, information processing, and transmission. All of them are pertinent in speculations about the beginnings of living systems.

References

Andrewartha, H. G., and Birch, L. C. 1954. *The distribution and abundance of animals*. Chicago: Univ. of Chicago Press.

Ashby, W. R. 1954. *Design for a brain*, 2d imp. London: Chapman & Hall.

————. 1956. *An introduction to cybernetics*. London: Chapman & Hall.

Beer, S. 1961. Below the twilight arch: A mythology of systems. In *Systems: Research and design*, ed. D. P. Eckmann, pp. 1–25. New York: Wiley.

Bodenheimer, F. S. 1958. *Animal ecology today*. Uit. W. Junk, Den Haag.

Bray, J. R. 1960. The chlorophyll content of some native and managed plant communities in central Minnesota. *Canad. J. Bot.* 38:313–33.

Buchanan, J. B. 1964. A comparative study of some features of the biology of *Amphiuma filiformis* and *Amphiuma chiajei* (*Ophiuroidea*) considered in relation to their distribution. *J. Mar. Biol. Ass. U.K.* 44:565–76.

Clark, H. F., and Bourlière, F. 1963. *African ecology and human evolution*. Chicago: Aldine.

Clements, F. E. 1920. *Plant indicators*. Carnegie Institute Publication No. 290. Washington, D.C.: Carnegie Inst.

Connell, J. H., and Orias, E. 1964. The ecological regulation of species diversity. *Amer. Nat.* 98:399–414.

Cowles, H. C. 1899. The ecological relations of the vegetation on the sand dunes of Lake Michigan. *Botan. Gaz.* 27:95–117, 167–202, 281–308, 361–91.

Crisp, D. J. 1962. Swarming of planktonic organisms. *Nature* 193:597–98.

Dansereau, P. 1954. Climax vegetation and the regional shift of controls. *Ecology* 35:575–79.

Erdman, G. 1963. Palynology and pleistocene ecology. In *North Atlantic biota and their history*, pp. 367–75. Oxford: Pergamon.

Florkin, M. 1949. *Biochemical evolution*. New York: Academic.

Fokker, A. D. 1966. Body and soul in the light of physics. *Proc. Konink. Nederl. Akad. Vetens.* (ser. B) 69:319–26.

Friedrich, H. 1955. Materialien zur Frage der Artbildung in der Fauna des marinen Pelagials. *Veroff. Inst. Meeresf. Bremerhaven* 3:159–89.

Haldane, J. B. S. 1957. The cost of natural selection. *J. Genet.* 55:511–24.

Herrera, J., and Margalef, R. 1963. Hidrografía y fitoplancton de la costa comprendida entre Castellon y la desembocadura del Ebro, de julio de 1960 a junio de 1961. *Inv. Pesq.* 24:33–112.

Hessler, R., and Sanders, H. 1966. The diversity of the benthic fauna of the sea. *II. Intern. Congr. Oceanography* (abstracts), pp. 157–58.

Huffaker, C. B. 1958. Experimental studies on predation: Dispersion factors and predator-prey oscillations. *Hilgardia* 27:343–83.

Kerner, E. H. 1957. A statistical mechanics of interacting biological species. *Bull. Math. Biophys.* 19:121–46.

———. 1959. Further considerations on the statistical mechanics of biological associations. *Bull. Math. Biophys.* 21:217–55.

Kershaw, K. A. 1963. Pattern in vegetation and its causality. *Ecology* 44:377–88.

Kurten, B. 1953. On the variation and population dynamics of fossil and recent mammal populations. *Acta Zool. Fennica* 76:1–122.

Landauer, R. 1961. Symposium on self-regulation in living systems. *Nature* 189:800.

Leeuwen, C. G. van. 1965. Het verband tussen natuurlijke en anthropogene landschapsvormen, bezien vanuit de betrekkingen in grensmilieu's. *Gorteria* 2:93–105.

Leigh, E. G. 1965. On the relation between the productivity, biomass, diversity, and stability of a community. *Proc. Nat. Acad. Sci.* 53:777–83.

Levins, R. 1962–63. Theory of fitness in a heterogeneous environment, I, II. *Amer. Nat.* 96:361–73, 97:75–89.

Lotka, A. J. 1956. *Elements of mathematical biology*. New York: Dover.

MacArthur, R. H. 1955. Fluctuations of animal populations, and a measure of community stability. *Ecology* 36:533–36.

Margalef, R. 1958. Temporal succession and spatial heterogeneity in natural phytoplankton. In *Perspectives in marine biology*, pp. 323–49. Berkeley and Los Angeles: Univ. of California Press.

Margalef, R. 1959. Mode of evolution of species in relation to their places in ecological succession. *XV Intern. Congr. Zool.* (London), section X, paper 17.

———. 1960. Valeur indicatrice de la composition des pigments du phytoplancton sur la productivité, composition taxonomique et propriétés dynamiques des populations. *Rapp. Proc.-verb. C. I. E. S. M. M.* 15:277–81.

———. 1961a. Communication of structure in planktonic populations. *Limol. Oceanogr.* 6:124–28.

———. 1961b. Corrélations entre certains charactères synthétiques des populations de phytoplancton. *Hydrobiologia* 18:155–64.

———. 1962a. "Diversita" dello zooplancton nel lago Maggiore. *Mem. Ist. Ital. Idrobiol.* 15:137–51.

———. 1962b. Succession in marine populations. *Advancing Frontiers of Plant Sciences* 2:137–88. Institute for the Advancement of Science and Culture, New Delhi.

———. 1963a. Algunas regularidades en la distribucion a escala pequeña y media de las poblaciones marinas de fitoplancton y del valor indicador de sus pigmentos. *Inv. Pesq.* 23:11–52.

———. 1963b. On certain unifying principles in ecology. *Amer. Nat.* 97:357–74.

———. 1964. Correspondence between the classic types of lakes and the structural and dynamic properties of their populations. *Verh. Intern. Verein. Limnol.* 15:169–75.

———. 1965a. Diversidad de las muestras de poblaciones de peces en funcion de la madurez del ecosistema, de la intensidad de explotacion y de la selectividad de las artes. *V. Reunion sobre Productividad y Pesquerias*, pp. 113–15. Barcelona: Inst. Inv. Pesqueras.

———. 1965b. Ecological correlations and the relationship between primary productivity and community structure. *Mem. Ist. Ital. Idrob.*, suppl. 18, pp. 355–64.

———. 1966. Análisis y valor indicador de las comunidades de fitoplancton mediterraneo. *Inv. Pesq.* 30:429–82.

———. 1967a. The food-web in the pelagic environment. *Helgolander Wiss. Meeresunters* 18:548–59.

———. 1967b. Some concepts relative to the organization of plankton. *Oceanogr. Mar. Biol. Ann. Rev.* 5:257–89.

Margalef, R., and Herrera, J. 1963. Hidrografia y fitoplancton de las costas de Castellon, de julio de 1959, a junio de 1961. *Inv. Pesq.* 22:49–109.

Margalef, R., Herrera, J., Steyaert, M., and Steyaert, J. 1966. Distribution et caractéristiques des communautes phytoplanctoniques dans le bassin tyrrhenien de la Méditerranée en fonction des facteurs ambiants et à la fin de la stratification estivale de l'année 1963. *Bull. Inst. Roy. Sci. Nat. Belgique* 42(5):1–56.

Mathew, W. D. 1915. Climate and evolution. *Ann. N. Y. Acad. Sci.* 24:271–318.

Needham, A. E. 1959. The origination of life. *Quart. Rev. Biol.* 34:189–209.

Patten, B. C. 1961. Competitive exclusion. *Science* 134:1599–601.

Pielou, E. C. 1966. Species-diversity and pattern-diversity in the study of ecological succession. *J. Theoret. Biol.* 10:370–83.

Pimentel, D. 1961. Species diversity and insect population outbreaks. *Ann. Ent. Soc. Amer.* 54:76–86.

Platt, J. R. 1961. "Bioconvection patterns" in cultures of free swimming organisms. *Science* 134:1766–767.

Platt, R. B. 1965. Ionizing radiation and homeostasis of ecosystems. In *Ecological effects of nuclear war*, pp. 39–60. Upton, N.Y.: Brookhaven Nat. Lab.

Shelford, V. E. 1963. *The ecology of North America*. Urbana, Ill.: Univ. of Illinois Press.

Slobodkin, L. B. 1961. *Growth and regulation of animal populations*. New York: Holt, Rinehart and Winston.

Tanaka, O., Irie, H., Izuka, S., and Koga, F. 1961. The fundamental investigation on the biological productivity in the North-West of Kyushu. I. The investigation of plankton. *Rec. Oceanogr. W. Japan*, special no. 5, pp. 1–57.

Taylor, G. 1934. The ecological basis of anthropology. *Ecology* 15:223–42.

Thorson, G. 1957. Bottom communities (sublittoral or shallow shelf). *Geol. Soc. Amer. Mem.* 67:461–534.

Vives, F., and López-Benito, M. 1957. El fitoplancton de la Ria de Vigo, desde julio de 1955 a junio de 1956. *Inv. Pesq.* 10:45–146.

Volterra, V. 1926. Variazioni e fluttuazioni del numero d'individui in specie animali conviventi. *Mem. Accad. Lincei* (s. 6) 2:31–113.

Wiener, N. 1948. *Cybernetics*. New York: Wiley.

Index

Glucids, structural, utilization of, 44

Gradients in ecosystem: biomass, 22; diversity, 22; structure, 22, 69

Harvey units of pigment, 63

Heterogeneity, of ecosystems in space, 74

Homeostat, 57

Human. *See* Man

Imbalances, in succession and evolution, 94, 95. *See also* Accidents, historical

Increase, rates of, 9, 21

Information: content of nature, 2; in the definition of stability, 12; destruction, 9, 17; exchange, 17; preservation and accumulation, 5, 29; in the results of activity of organisms, 23; theory, 2. *See also* Channels, of information

Interactions, between species, 6, 7, 9

International Biological Programme (IBP), 50

Invasion, of sea by organisms of terrestrial origin, 94–95

Kinetophilous species, 88

Lakes: dystrophic, 43; entrophic, 41; oligotrophic, 42; succession in, 41

'Learning,' by the ecosystem, 29

Leaves, gradients in, 62

Leigh, E. G., on frequency of fluctuations, 12

Life: beginnings of, 100; span, 85–86; tables, 89

Loop. *See* Feedback

Loss. *See* Exploitation

Lotka, A. J., on interacting populations, 5, 10

MacArthur, R. H., concept of stability, 11

Macrophages, 76

Man: and culture evolution, 47; impact on nature, 96

Maps, 39, 70

Mathew, W. D., on climate and evolution, 96

Maturity: changes in, 37; of ecosystems, 33, 34, 91; and exploitation, 37; of lakes, 42–43; relative values of, 44

Metabolism, deficiencies of, 86

'Microscopic': aspects of ecosystem, 6, 23, 24; versus 'macroscopic,' 24

Miniaturization, in living structures, 3

Mortality, 10, 89, 90; age of maximum, 91

Natural selection: assumed wastefulness of, 81; cost of, 83; enlargement of concept, 3

Neoteny, 92, 95, 96

Nets, mesh size, 54

von Neumann's game, 30

Niche, 7, 8, 12, 31, 54

Noise, in ecological information channel, 21

Numbers of individuals, regularity in, 18

Organization, 14, 16; diversity as a measure of, 19; levels of, 4; spatial, 14, 15; subsystems of lower, 96

Orthogenesis, 27, 81, 89

Parthenogenesis, 92

Particle counters, for the study of plankton, 57–58

Patches, in the composition of communities, 41

Patterns, spatial, 20, 50

Phylogenetic series, 94

Pigments: diversity of, 58; as estimates of biomass, 63; as estimates